Frontiers in Economic History

Economic historians have contributed to the development of economics in a variety of ways, combining theory with quantitative methods, constructing new databases, promoting interdisciplinary approaches to historical topics, and using history as a lens to examine the long-term development of the economy. Frontiers in Economic History publishes manuscripts that push the frontiers of research in economic history in order to better explain past economic experiences and to understand how, why and when economic change occurs. Books in this series will highlight the value of economic history in shedding light on the ways in which economic factors influence growth as well as social and political developments. This series aims to establish a new standard of quality in the field while offering a global discussion forum toward a unified approach in the social sciences.

Jakub Growiec

Accelerating Economic Growth

Lessons From 200,000 Years of Technological Progress and Human Development

Springer

Jakub Growiec
Department of Quantitative Economics
SGH Warsaw School of Economics
Warsaw, Poland

ISSN 2662-9771 ISSN 2662-978X (electronic)
Frontiers in Economic History
ISBN 978-3-031-07194-2 ISBN 978-3-031-07195-9 (eBook)
https://doi.org/10.1007/978-3-031-07195-9

This Springer imprint is published by the registered company Springer Nature Switzerland AG
The registered company address is: Gewerbestrasse 11, 6330 Cham, Switzerland

For Antoni
may he live in an awesome world

Acknowledgements

I am grateful to, (in alphabetical order) Olivier Bargain, Jan Frankowski, Pedro Gil, Marek Góra, Michał Gradzewicz, Michał Lewandowski, Michał Opuchlik, Ingmar Schumacher, Andrzej Sławiński and Łukasz Woźny, for their useful suggestions and illuminating discussions.

Above all I greatly appreciate both the patience and support, and the inspiring conversations with my wife, Katarzyna Growiec. She's an excellent social psychologist working among other topics on social networks, social capital, and the spread of trust and cooperation in societies. These are important threads for this book as well.

My scientific work on the hardware–software model has been financially supported by the Polish National Science Center (Narodowe Centrum Nauki) under grant OPUS 14 No. 2017/27/B/HS4/00189 titled: "R&D Capital, Artificial Intelligence and Growth".

As opposed to my research papers where the official disclaimer tends to be a mere formality, this book actually presents strong views. Therefore it must be emphasized that these views are mine and mine only. They have not been endorsed by Narodowy Bank Polski or SGH Warsaw School of Economics, with which I am affiliated.

I solemnly apologize for any possible mistakes. They're all on me.

Contents

List of Figures

List of Tables

Chapter 1
Introduction

The consequences for human welfare involved in questions like these [i.e., regarding the pace of long-run economic growth – JG] are simply staggering: once one starts to think about them, it is hard to think about anything else.

Robert E. Lucas (1988), p. 5

1.1 What to Expect from This Book

This book aims to shed new light on the following questions: What drives technological progress and economic growth over the long run and at the global scale? What made it possible for the single human species to conquer the world, build a global digital economy and still want more (Fig. 1.1)? And how will technological progress, economic growth and the overall prosperity of human civilization unfold in the future?

These are very old and often addressed questions. The answers, however, are a mixed bag. On the one hand, the scholarly literature on economic growth can boast many successes in identifying proximate causes of growth, from the accumulation of physical and human capital to political institutions and cultural norms. It has also specified the causal mechanisms that explain how new ideas can spur lasting growth. The literature has also come a long way in describing and theorizing about the two key technological revolutions in human history, that initiated agriculture and industry. On the other hand, my experience with this topic also leads to a feeling of discontent with how partial and shaky our understanding of global growth processes still is.

First, there appears to be a growing dissonance between the documented properties of the contemporary global economy and the way in which it is traditionally analyzed in the macroeconomic literature, both empirically, e.g., in growth and

J. Growiec, *Accelerating Economic Growth*, Frontiers in Economic History,
https://doi.org/10.1007/978-3-031-07195-9_1

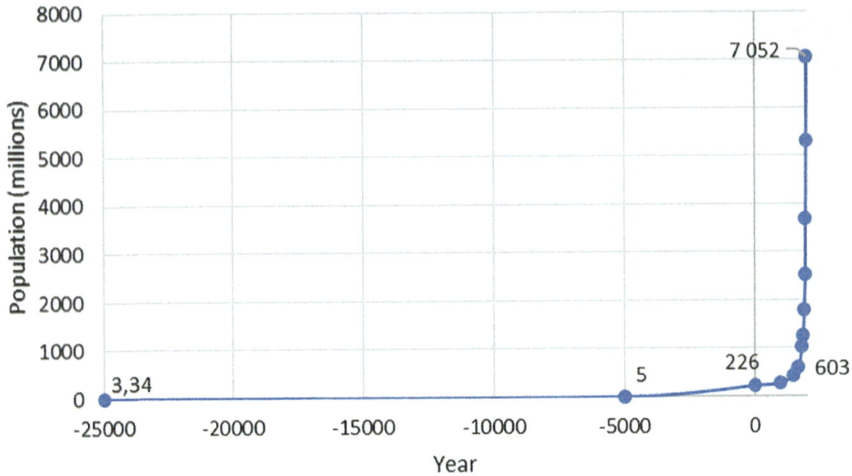

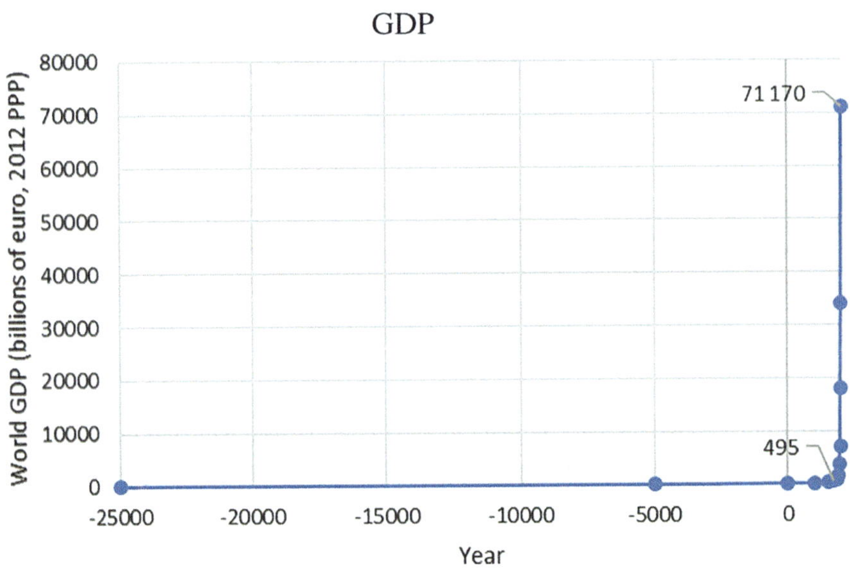

Fig. 1.1 World population and GDP, 25 000 BCE—2012 CE. *Notes* GDP per capita assumed constant until 0 CE. The values: 603 million people and 495 billions of euro, PPP-adjusted in 2012 prices, pertain to the year 1700 CE. *Sources* Kremer (1993a) for data points before 0 CE, Piketty (2014) otherwise

development accounting exercises based on National Accounts data (Jorgenson and Stiroh, 2000; Timmer and van Ark, 2005; Jorgenson, 2005; Fernald, 2015; Gordon, 2016; Cette et al., 2016), and theoretically, in formal models of economic growth and technological change (Romer, 1990; Kremer, 1993a; Jones, 1995, 1999; Ha and Howitt, 2007; Madsen, 2008; Bloom et al., 2020; Kruse-Andersen, 2017). Economics literature is still struggling to produce a convincing theoretical framework that would reconcile past growth experiences with emerging new global trends such as declining labor shares (Karabarbounis and Neiman, 2014), increasing profit shares (Barkai, 2020), increasing markups and market power (De Loecker and Eeckhout, 2017; Jones, 2018; Diez et al., 2018), increasing within-country income inequality (Piketty, 2014; Piketty and Zucman, 2014), increasing market concentration (Autor et al., 2017), skill polarization and the elimination of routine jobs (Acemoglu and Autor, 2011; Autor and Dorn, 2013), the overall increasing susceptibility of jobs to automation (Frey and Osborne, 2017; Arntz et al., 2016), and the fact that biggest fortunes these days are made in the software business.

Second, the actual scope of consequences of these new trends may be even underestimated because we're looking at the global economy through the lens of obsolete measurements and theories. The economy has become globally interdependent and propelled by the flow of digital information, whereas our thinking remains locked in times when it was, for the most part, country-specific and propelled by the flow of industrial commodities. Our picture of the world lags behind reality by about a generation (Rosling, 2018). It's about time to bridge that gap.

The new angle of this book consists in borrowing a few ideas from physics, anthropology, psychology, history, moral philosophy and the studies of artificial intelligence (AI), and connecting them to a brand new economic growth theory—the "hardware–software model". I take an a-historical and a-geographical approach. My challenge is to build an overarching theoretical structure that is valid universally, across all time and space. To this end, I deliberately move away from the usual "wealth of nations" perspective and, in particular, stop asking questions like why some countries are richer than others or why the Industrial Revolution was initiated in England rather than somewhere else. I also deliberately take an informed view of the past, equipped with modern concepts and notions, and all the positive knowledge that we have today. Instead of trying to understand the conscious motivations behind the actions taken by our ancestors, I just look at these actions and study the broad patterns that emerge from them. Offering a bird's eye view of past development processes, I intentionally disregard phenomena that appear only locally (like the phenomenal growth experience of "Asian tigers" in the late twentieth century) or that are short-lived (like business cycles).

In Part I of this book I argue that human expansion, technological progress and economic growth have the same fundamental driver: the mechanism of *local control maximization*. This mechanism emerges as a side effect of human actions and thus need not be pursued consciously. It is a sub-routine of the grand process of the evolution of species, which got out of hand in the case of humans because of its pace (faster by orders of magnitude than the pace of evolution), recursive self-improvement and greed for resources.

Fig. 1.2 The giant pendulum of global economic development

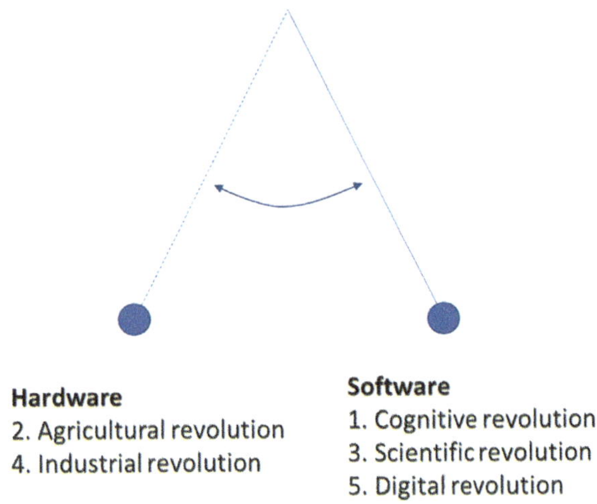

With this development drive working in the background, I organize the millennia of human history into five eras of technological and economic development: the hunter–gatherer era, the agricultural era, the enlightenment era, the industrial era and the digital era (Galor, 2011; Harari, 2014). The eras have been initiated by five respective technological revolutions: (i) the cognitive (Upper Paleolithic) revolution (ca. 70 000 BP)—the time when *homo sapiens* acquired the status of a dominant hominin species and began gaining control over natural habitats across the world; (ii) the agricultural (Neolithic) revolution (ca. 10 000 BP), marking the beginning of transition from hunting and gathering to sedentary agriculture; (iii) the scientific revolution (ca. 1550 CE) that initiated modern science; (iv) the industrial revolution (ca. 1800 CE), the onset of industrial production, capital accumulation and systematic increases in GDP per capita; and (v) the digital revolution (ca. 1980 CE) that initiated a rapid explosion in the world's capacity to compute, store and communicate data.

Naturally, this list of revolutions is subject to scientific dispute and their dating is at best provisional. For example, there were probably seven independent agricultural revolutions at seven different dates (Diamond, 1997); the approximate date of 10 000 years ago marks the first emergence of agriculture in the Fertile Crescent (in the Middle East) but not the later ones that appeared elsewhere. Nevertheless, there is one intriguing regularity that I exploit in Part II of this book which makes the above list of five eras a viable candidate. Namely, the consecutive revolutions affected, in an alternating fashion, *hardware* (technologies of performing physical action) and *software* (technologies of processing information). The cognitive, scientific and digital revolutions affected software, whereas the agricultural and industrial revolutions affected hardware. The history of human technology, economy and society was like a giant pendulum, swinging back and forth between the primacy of hardware and software (Fig. 1.2) as the key domain of development.

In Part III of this book I emphasize that each new era opens an entirely *new dimension of development*, operating on top of earlier ones and exhibiting much faster (factually, orders of magnitude faster) growth, and therefore gradually gaining dominance over previous-era activities. For example, the agricultural revolution introduced the possibility of increasing production of food per one unit of land area, thereby greatly boosting population density. The industrial revolution, in turn, initiated energy generation from fossil fuels and large-scale accumulation of productive capital, thus marking the beginning of unprecedented growth in output per worker and standards of living of the general population.

At any given moment in time, however, technologies from different eras coexist and—if you are interested in the "wealth of nations" perspective—constitute the prime source of cross-country inequality in income and wealth. Countries with strong research and ICT sectors tend to be richer than countries specializing in industrial production, which are themselves richer than countries specializing in agriculture. The wealthiest people in the world are owners of US software companies, whereas in the poorest regions of Sub-Saharan Africa most people are still engaged in subsistence agriculture. Furthermore, to complicate the matters somewhat, there are strong feedback loop effects across eras so that developments of the new era reinforce the previous-era economy. For example, developments of the scientific and industrial revolutions such as tractors, fertilizers and new plant varieties massively boosted agricultural productivity, whereas computers with broadband Internet brought major cost reductions across the manufacturing industry.

Finally, Part IV of this book is concerned with the future. Even though "it is very difficult to predict, especially the future" (as Niels Bohr used to say), I provide some new value in this regard by looking at the unfolding digital era from a global, long-run perspective, using my hardware–software economic growth model as a guide for identifying the decisive trends. The presence of systematic patterns linking the consecutive technological and economic revolutions underscores that viewing the digital era as a mere continuation of the industrial era is a really bad idea. In our predictions we must not miss such emerging new trends as the quick decoupling of processing, storage and communication of information from the cognitive capacities of the human brain, the growing scale mismatch between the economy (full of global value chains and international trade linkages) and policy (which remains stubbornly national), systematic correlation between the spread of highly scalable digital technologies and increasing income inequality, or the mounting cumulative progress towards stronger and more general AI algorithms.

In sum, there are three distinct thoughts in this book that are genuinely new: (i) local control, (ii) the hardware–software model, and (iii) the idea that technological revolutions open up new dimensions of development. It's also the first book that takes these pieces of a puzzle together—as well as a few additional ones—and uses them in a theory of global economic growth and technological progress over the entire human history, with intriguing new predictions for the future.

1.2 Customers Who Bought This Item Also Bought

Although I am an academic economist and my ideas are inspired primarily by the scholarly literature on economic growth, this book was also influenced by a few excellent monographs and essays (as well as, I should say, blog posts and YouTube videos) outside of economics. My foremost inspiration was the fascinatingly broad narrative on the origins of agriculture, industry and technological advances: *Guns, Germs and Steel* by the anthropologist, ecologist and geographer (Diamond, 1997). My second inspiration was another celebrity monograph: *Sapiens. A Brief History of Humankind* by the historian (Harari, 2014). If Diamond's book is a journey, Harari's is a roller-coaster—filled with inspiring stories and open questions for the future. I doubt if it is possible to tell the story of humankind better than these two outstanding authors did. Yet, as no story is ever fully complete, I've decided to build on their insights and expand them. Indeed, my ambition—to build a synthetic theory of economic growth and technological progress over the entire human history, framing the millennia of past development as a single cumulative process—is pretty much what these books are doing, too. So if you say my book just adds footnotes to these two epic writings, I'd be happy to accept that view. I just hope these footnotes add some new value.

Within economics, readers interested in the long span of economic growth should appreciate Oded Galor's monograph *Unified Growth Theory* (Galor, 2011). It says much more than I can ever say about the origins of agriculture and industry, and about the dynamics and international disparities of population, technology and economic growth.

There is also a number of books that document important new global trends unveiling in the digital era. Thomas Piketty's *Capital in the Twenty-First Century* (Piketty, 2014)—even though I disagree with some of his provocative claims—presents extremely valuable new data on the functional distribution of income. His observation that the labor's share of income is declining alongside with the increasing income inequality within societies is certainly very real. Erik Brynjolfsson and Andrew McAfee's *The Second Machine Age* (Brynjolfsson and McAfee, 2014) demonstrates the dormant power of new digital technologies that may one day be unleashed, bringing about thrilling new opportunities as well as mounting risks, including automation of more and more middle-class jobs and further increases in inequality. Robert Gordon's *The Rise and Fall of American Growth* (Gordon, 2016) provides a valuable discussion of the flipside of Brynjolfsson and McAfee's enthusiasm, showing us that US data can also be viewed as an indication of an upcoming secular growth slowdown: apparently, measured returns to digital technologies are declining. All these books, alongside a long list of related scholarly papers, indicate that the modeling apparatus that economists have perfected for describing and predicting the industrial economy may be no longer sufficient. New facts undermine the conviction that the industrial economy is the final stage of development, following the earlier hunter–gatherer and agricultural eras. On the contrary, the digital era is qualitatively different, and recent growth has in fact been quite unbalanced in certain important ways.

Finally, I was also strongly inspired by the literature on artificial intelligence, in particular the essays on the promises and perils of potential appearance of artificial general intelligence (AGI), i.e., algorithms able to optimize their objective functions more efficiently than human brains do, including the ability to act adaptively and apply their optimization skills across a wide range of different domains. Nick Bostrom's *Superintelligence: Paths, Dangers, Strategies* (Bostrom, 2014) was a key inspiration in this regard, offering the instrumental convergence thesis that I extrapolate back from AI algorithms to the biological world, including humans (see also Bostrom, 2012). The same author also devised the "paperclip maximizer" AGI example (Bostrom, 2003), perhaps the most disappointing of all plausible doomsday scenarios (on par with the "grey goo" nanobot scenario, go figure). The best place to get really excited about this topic is, however, Tim Urban's two outstanding blog posts on waitbutwhy.com about AI and one about brain–machine interfaces. (I'm all in the Anxious Avenue.) I was also stunned by various writings of Eliezer Yudkowsky on lesswrong.com, as well as his essay *Intelligence Explosion Microeconomics* (Yudkowsky, 2013) and his debate with Robin Hanson on the risks of hard AI takeoff, *AI Foom* (Hanson and Yudkowsky, 2013). With all respect to Hanson and his brilliant ideas, I firmly side with Yudkowsky in this debate.

1.3 Five Eras, Initiated by Five Revolutions

The narrative of long-run economic growth and technological progress over the millennia of human history can be intuitively organized as five eras, initiated by five consecutive technological revolutions, such that each new era puts development into a higher gear. Let me begin with (Y-chromosomal) Adam and (mitochondrial) Eve.

Hominins first appeared on the face of Earth about 2 400 000 years ago in East Africa (Dunsworth, 2010). According to the (arguably most popular) Out-of-Africa theory, the same geographic region was also the cradle of the anatomically modern human, *homo sapiens* (Ashraf and Galor, 2013). Our first ancestors emerged in the African savannas around 200 000 years before present (BP). (The exact timing is uncertain, though. New archeological discoveries in Morocco (Hublin et al., 2017) suggest that the *Homo sapiens* is not 200 000, but more than 300 000 years old.) Apart from establishing their presence in the region, in the course of tens of thousands of years they also made a few early unsuccessful efforts to expand their habitats toward Asia and Europe.

The first turning point in human history was the *cognitive revolution*, also called the Upper Paleolithic revolution. Its dating is highly uncertain; for orientation let us place it around 70 000 BP (Harari, 2014). Newly acquired cognitive skills, such as advanced language and an enhanced theory of mind, allowed humans to become not only anatomically but also behaviorally modern and advance from the middle to the top of the food chain (Tattersall, 2009; Tomasello, 2014; Harari, 2014). What followed was a wave of human dispersal out of Africa which eventually turned out permanent. The expansion was probably originated by as few as 1000 individuals

(Liu et al., 2006). It followed a "southern route" via the Bab al-Mandab strait toward South Asia and Australia, which was reached about 65 000 BP–50 000 BP. Europe, in contrast, was colonized by the *homo sapiens* much later, only about 45 000 BP, despite geographical proximity and lack of physical barriers. The hypothesized reason for this delay is that at the time, Europe was populated by the *homo neanderthalensis*. This coincidence signifies that Europe's Neanderthals were an ecological barrier for modern humans, who could only enter Europe when the demise of Neanderthals had already started.

The *hunter–gatherer era* that followed the cognitive revolution constituted the first major step of development that has been achieved by humankind but has never been achieved by any other species. The perspective of other species elucidates how impressive it was that human hunter–gatherer bands and tribes managed to colonize almost all habitats in the world, including passing the Bering Strait to the Americas and inhabiting most Pacific archipelagos. And while they were taking successive natural habitats under their control, they drove all other hominins to extinction-and did the same for a wide variety of other species as well, such as the Asian mammoths, Australian megafauna including huge diprotodons (Harari, 2014), or American saber-toothed cats.

However, by today's standards, they only managed to obtain minuscule "economic growth" rates. It took the *homo sapiens* whopping 20–25 millennia to defeat the Neanderthals and colonize Europe. Around 7 000 BP (5 000 BCE) the combined human population of the world was in the order of just 5 million (Kremer, 1993a). Yet, this pace of development was already at least an order of magnitude faster compared to the pace of the evolution of species.

The second turning point in human history was the *agricultural revolution*, also called the Neolithic revolution (Diamond, 1997; Hibbs and Olsson, 2004; Bocquet-Appel, 2011). Domestication of plants and animals allowed for dramatic increases in the amount of calories derived from a given land area, leading to marked increases in human population density, and encouraging the roving hunter–gatherer tribes to abandon their former lifestyle in favor of sedentary agriculture (Diamond, 1997; Hibbs and Olsson, 2004).

The *agricultural era* was initiated ca. 10 000 BP (8 000 BCE) in the Fertile Crescent (in the Middle East). The first eight "Neolithic founder crops" included wheat, barley, lentil, pea and chickpea. Independent agricultural revolutions, with different founder crops, appeared subsequently over later millennia in various other parts of the world: Central China, the New Guinea Highlands, Central Mexico, northwestern South America, Sub-Saharan Africa and eastern North America. Adoption of agriculture was only gradual because gains in food production were initially modest and technological progress as well as diffusion was slow. For these reasons, agricultural societies coexisted with roving hunter–gatherer tribes for entire millennia before the former managed to outnumber and thus economically dominate the latter (Diamond, 1997). In fact, some populations never managed to shift to sedentary agriculture, e.g. the Aboriginal Australians before the European conquest.

Agricultural-era technologies and economies were developed for millennia, and it is of course a very rough approximation to treat this period as homogeneous.

The agricultural era witnessed not only a gradual buildup of population density and conversion of natural ecosystems into farmland, but also the creation of centralized states with increasing capacity (Borcan et al., 2018) as well as the rise and fall of empires, notably the Roman empire, Imperial China under the Han dynasty, empires of the Mongols, Incas, etc. The invention of writing, that first appeared around 3 400 BCE in ancient Sumer, allowed for lasting and reliable storage of information and facilitated its spread. Despite all the progress, however, the agricultural period also included a few episodes of population and economic decline (e.g., following the fall of the Western Roman empire in 476 CE or during the Mongol conquest of China in the thirteenth century). But the overall global trend of technological development and economic growth was nevertheless systematically positive.

The third turning point in human history was the European *scientific revolution* of the sixteenth century CE. In this period, diffusion and storage of knowledge and information was facilitated, thanks to the rapid spread of Johannes Gutenberg's (or Laurens Coster's?) fifteenth-century invention of the printing press. It was also the time of great geographical discoveries and European colonial conquests.

These developments lay the foundations for the creation of modern science, including empiricism, experimentation, mathematization and hypothesis testing. Europe began to build universities and witnessed a blooming of pioneering and belief-twisting works by Nicolaus Copernicus, Galileo Galilei, Johannes Kepler, Isaac Newton, Blaise Pascal, Gottfried Leibniz and others. This all resulted in the arrival of the *enlightenment era*, culminating in the 18th century. Enlightenment not only granted humankind with brilliant new ideas but also—crucially—with an environment that facilitated the nurturing, spreading and further development of such ideas. Never before has philosophy and science been so successful in explaining the world and opening new development opportunities.

The fourth turning point in human history was the *industrial revolution* (Galor and Weil, 2000; Hansen and Prescott, 2002; Galor, 2005, 2011). It began around 1800 CE, first in England, and then it quickly spread across Europe and into the Western Offshoots such as the US. Formerly agricultural societies began to harness energy from fossil fuels, build factories, manufactures and engage in mass production. The set of key technological inventions which fueled this shift included, among many others, steam engine, electricity, internal combustion engine and indoor plumbing (Gordon, 2016). The industrial revolution was critically enabled by the earlier scientific revolution. It initiated systematic increases in output per capita and growth in demand for skilled labor, activated the hitherto dormant children quantity—quality tradeoff, and by lowering fertility broke the Malthusian link between development and population size (Galor, 2011). The vicious circle in which increased prosperity inevitably led to a larger population, which in turn reduced per capita wealth back to the subsistence level, finally gave way, allowing for broad-based increases in wealth and standards of living.

As opposed to the agricultural revolution, the industrial revolution happened only once in history. At the time, the world was already connected, and the gap between technological knowledge acquired by the Western societies and any other disconnected society was too large to enable an independent industrial revolution.

The technologies and economic institutions of the *industrial era* were gradually adopted across the world, increasing the humankind's economic power and enabling acceleration of economic growth rates by an order of magnitude: doubling times of global output were cut from hundreds of years to just decades (Galor, 2005). The transition to the industrial era is not yet complete, though: subsistence agriculture still employs a majority of population in parts of Sub-Saharan Africa or India. Major pockets of agricultural poverty can also be found across, e.g., China, Indonesia, the Middle East or Latin America.

Before the fruit of the industrial era has been fully reaped, the humankind entered yet another technological revolution: the *digital revolution*. A tentative timing would place this revolution around the 1980s when personal computers have begun to permeate firms and households; however, the revolution really gained momentum in the 2000s when the Internet connected the computers to a global World Wide Web.

The key inventions which fueled the digital revolution were the Turing machine, semi-conductors and integrated circuits, followed by what now is the hallmark of the *digital era*: personal computers, the Internet, cell phones, smartphones and industrial robots. First key developments were achieved (mostly) in the US, but their diffusion was really fast and nowadays a major part of the hardware frontier of the digital economy has moved to East Asia (Japan, Korea, Taiwan and China). On the consumer side, even very poor and infrastructurally disadvantaged parts of the world frequently use (at least some) digital technologies. An illustrative statistic describing the pace of development in the digital era is Moore's Law (Hilbert and López, 2011): since the 1980s, the world's general-purpose computing capacity doubles every two years.

Looking from 2022 into the future, we cannot exclude the possibility of further technological revolutions. As argued by some authors, such a revolution may likely occur already within the current century (Hanson, 2000; Kurzweil, 2005; Brynjolfsson and McAfee, 2014), with totally unpredictable consequences (perhaps unless you're into sci-fi, Lem, 1964). This gets only more intriguing given that the digital-era technology is already maturing: maintaining the pace of Moore's Law requires employing more and more resources (Bloom et al., 2020) and the scope for further miniaturization (fitting more transistors on a microprocessor chip) appears to be already exhausted (Waldrop, 2016).

Focusing only on the population which resides in a given place on the world map, one may get the impression that the humankind developed linearly, gradually passing from the hunter–gatherer to the agricultural, enlightenment, industrial and digital eras. However, this wasn't actually the case because (i) the timing of take-off to the next era varied largely across the world (Diamond, 1997; Galor, 2005), (ii) the take-off was only gradual (and therefore the technologies from consecutive eras coexist not only globally but also locally, and between-era inequality is typically far greater than within-era inequality) and (iii) each new era strongly feeds back on the previous-era economy: just think of the gains from mechanization and specialization of agriculture, or the computerization and robotization of industry.

Four main tendencies emerge from this narrative, linking the prehistoric hunter–gatherer times with the modern digital world. First, each new era accelerates "economic growth" by at least an order of magnitude. Second, each new era opens up an entirely new avenue of development, thus redefining what "economic growth" refers to: the capacity of natural habitats under human control, global population size, global GDP, or perhaps the global volume of processed and communicated data. Third, there is a curious alternating pattern such that if the previous revolution affected *hardware* (ways of doing physical stuff) then the next one affects *software* (ways of processing information), and vice versa. Fourth, the arrival of a new era does not end the previous era, instead the technologies from the consecutive eras coincide in time.

References

Acemoglu, D., & Autor, D. (2011). Skills, tasks and technologies: Implications for employment and earnings. In O. Ashenfelter & D. Card (Eds.), *Handbook of labor economics* (Vol. 4, Chap. 12, pp. 1043–1171).

Arntz, M., Gregory, T., & Zierahn, U. (2016). *The risk of automation for jobs in OECD countries: A comparative analysis, OECD social, employment and migration* (Working Paper No 189). OECD Publishing.

Ashraf, Q., & Galor, O. (2013). The out of Africa hypothesis, human genetic diversity, and comparative economic development. *American Economic Review, 103*, 1–46.

Autor, D. H., & Dorn, D. (2013). The growth of low-skill service jobs and the polarization of the US labor market. *American Economic Review, 103*(5), 1553–1597.

Autor, D., Dorn, D., Katz, L. F., Patterson, C., & Van Reenen, J. (2017). *The fall of the labor share and the rise of superstar firms* (Working Paper No. 23396). NBER.

Barkai, S. (2020). Declining labor and capital shares. *Journal of Finance, 75*, 2421–2463.

Bloom, N., Jones, C. I., Van Reenen, J., & Webb, M. (2020). Are ideas getting harder to find? *American Economic Review, 110*(4), 1104–1144.

Bocquet-Appel, J.-P. (2011). When the world's population took off: The springboard of the neolithic demographic transition. *Science, 333*, 560–561.

Borcan, O., Olsson, O., & Putterman, L. (2018). State history and economic development: Evidence from six millennia. *Journal of Economic Growth, 23*, 1–40.

Bostrom, N. (2003). Ethical issues in advanced artificial intelligence. In G. E. Lasker, W. Wallach, & I. Smit (Eds.), *Cognitive, emotive and ethical aspects of decision making in humans and in artificial intelligence* (Vol. 2). International Institute of Advanced Studies in Systems Research and Cybernetics.

Bostrom, N. (2012). The superintelligent will: Motivation and instrumental rationality in advanced artificial agents. *Minds and Machines, 22*, 71–85.

Bostrom, N. (2014). *Superintelligence: paths, dangers, strategies*. Oxford University Press.

Brynjolfsson, E., & McAfee, A. (2014). *The second machine age: Work, progress, and prosperity in a time of brilliant technologies*. Norton & Co.

Cette, G., Lecat, R., & Ly-Marin, C. (2016). *Long-term growth and productivity projections in advanced countries* (Working Paper No. 617). Banque de France.

De Loecker, J., & Eeckhout, J. (2018). *Global market power* (Working Paper 24768). National Bureau of Economic Research.

De Loecker, J., & Eeckhout, J. (2017). *The rise of market power and the macroeconomic implications* (Working Paper 23687). National Bureau of Economic Research.

Diamond, J. (1997). *Guns, germs, and steel: The fates of human societies*. Norton & Company.

Diez, F., Leigh, D., & Tambunlertchai, S. (2018). *Global market power and its macroeconomic implications* (Working Paper No. 18/137). International Monetary Fund.

Dunsworth, H. M. (2010). Origin of the genus homo. *Evolution: Education and Outreach, 3*, 247.

Fernald, J. G. (2015). Productivity and potential output before, during, and after the Great Recession. *NBER Macroeconomics Annual,29*, 1–51.
Frey, C. B., & Osborne, M. (2017). The future of employment: How susceptible are jobs to computerisation? *Technological Forecasting and Social Change,114*, 254–280.
Galor, O. (2005). From stagnation to growth: unified growth theory. In P. Aghion & S. N. Durlauf (Eds.), *Handbook of economic growth* (pp. 171–293).
Galor, O. (2011). *Unified growth theory*. Princeton University Press.
Galor, O., & Weil, D. N. (2000). Population, technology, and growth: From Malthusian stagnation to the demographic transition and beyond. *American Economic Review,90*, 806–828.
Gordon, R. J. (2016): *The Rise and Fall of American Growth: The U.S. Standard of Living since the Civil War*. Princeton University Press.
Ha, J., & Howitt, P. (2007). Accounting for trends in productivity and R&D: A Schumpeterian critique of semi-endogenous growth theory. *Journal of Money, Credit and Banking,39*(4), 733–774.
Hansen, G. D., & Prescott, E. C. (2002). Malthus to Solow. *American Economic Review,92*, 1205–1217.
Hanson, R. (2000). *Long-term growth as a sequence of exponential modes* (Working paper). George Mason University.
Hanson, R., & Yudkowsky, E. (2013). *The Hanson-Yudkowsky AI-Foom Debate*. Machine Intelligence Research Institute.
Harari, Y. N. (2014). *Sapiens: A brief history of humankind*. Vintage.
Hibbs, D. A., & Olsson, O. (2004). Geography, biogeography, and why some countries are rich and others are poor. *Proceedings of the National Academy of Sciences,101*, 3715–3720.
Hilbert, M., & López, P. (2011). The world's technological capacity to store, communicate, and compute information. *Science,332*, 60–65.
Hublin, J. J., Ben-Ncer, A., Bailey, S., et al. (2017). New fossils from Jebel Irhoud, Morocco and the Pan-African origin of homo sapiens. *Nature,546*, 289–292.
Jones, C. I. (1995). R&D-based models of economic growth. *Journal of Political Economy,103*, 759–84.
Jones, C. I. (1999). Growth: With or without scale effects? *American Economic Review,89*(2), 139–144.
Jorgenson, D. W. (2005). Accounting for growth in the information age. In P. Aghion & S. Durlauf (Eds.), *Handbook of economic growth*.
Jorgenson, D. W., & Stiroh, K. J. (2000). Raising the speed limit: U.S. economic growth in the information age. *Brookings Papers on Economic Activity,1*, 125–211.
Karabarbounis, L., & Neiman, B. (2014). The global decline of the labor share. *Quarterly Journal of Economics,129*(1), 61–103.
Kremer, M. (1993). Population growth and technological change: One million B.C. to 1990. *Quarterly Journal of Economics,108*, 681–716.
Kruse-Andersen, P. K. (2017). *Testing R&D-based endogenous growth models* (Working paper). University of Copenhagen.
Kurzweil, R. (2005). *The singularity is near*. Penguin.
Lem, S. (1964). *Summa technologiae*. Wydawnictwo Literackie.
Liu, H., Prugnolle, F., Manica, A., & Balloux, F. (2006). A geographically explicit genetic model of worldwide human-settlement history. *American Journal of Human Genetics,79*, 230–237.
Madsen, J. (2008). Semi-endogenous versus Schumpeterian growth models: Testing the knowledge production function using international data. *Journal of Economic Growth,13*, 1–26.
Piketty, T. (2014). *Capital in the twenty-first century*. Translated from French by Arthur Goldhammer: Harvard University Press.
Piketty, T., & Zucman, G. (2014). Capital is back: Wealth-income ratios in rich countries, 1700–2010. *Quarterly Journal of Economics,129*(3), 1255–1310.
Romer, P. M. (1990). Endogenous technological change. *Journal of Political Economy,98*, S71–S102.

Rosling, H. (2018). *Factfulness: Ten reasons we're wrong about the world—and why things are better than you think*. Sceptre.

Tattersall, I. (2009). Human origins: Out of Africa. *Proceedings of the National Academy of Sciences,106*, 16018–16021.

Timmer, M. P., & van Ark, B. (2005). IT in the European union: A driver of productivity divergence? *Oxford Economic Papers,57*(4), 693–716.

Tomasello, M. (2014). *A natural history of human thinking*. Harvard University Press.

Waldrop, M. M. (2016). The chips are down for Moore's law. *Nature,530*, 144–147.

Yudkowsky, E. (2013). *Intelligence explosion microeconomics* (Technical report 2013-1). Machine Intelligence Research Institute.

Part I
The Drive for Development

Chapter 2
Enter Local Control

The purpose of life is to hydrogenize carbon dioxide.

attributed by Sean Carroll (2016) to Michael Russell

2.1 Evolution: Brought Us to Life and Left All Alone

Looking back at all the impressive achievements of humankind, all the technological revolutions, each of which constituted a major leap of human capability, begs the question: Why did mankind achieve it? What was the driving force behind all these developments?

A correct but only partial answer is species evolution. Arguably, in the beginning we were just lucky. Having tested plenty of other genetic designs of species, the evolutionary process somehow happened to produce the behaviorally modern *homo sapiens* who slightly exceeded other hominin species in terms of frontal cortex capacity of the brain (Harari, 2014)—and this slight intelligence advantage turned out sufficient for the human to go and rule the world.

Being "lucky" in the context of evolution underscores that the evolutionary process is not intentional when it explores the space of designs of species and maximizes genetic fitness of the surviving ones (commonly known as "survival of the fittest"). Its search involves a lot of randomness and is multidimensional so that species tend to fit into certain ecological niches by developing unique advantages (Darwin, 1859). It is also extremely slow: the fitness of innovations (mutations) is evaluated only *ex post* and over hundreds of generations. For large mammals like us, this means that major useful innovations are embodied in us at the time scale of hundreds of thousands of years. Hence, although species evolution never stops, we are anatomically indistinguishable from our ancestors who inhabited the Ethiopian savanna immediately after the cognitive revolution about 70 000 years ago (Ashraf and Galor, 2013).

J. Growiec, *Accelerating Economic Growth*, Frontiers in Economic History,
https://doi.org/10.1007/978-3-031-07195-9_2

A key reason why evolution cannot be more than a partial answer to our grand question is that ever since the cognitive revolution, the evolution of species just doesn't keep up with the pace of development of the human civilization. It was left behind by our early hunter–gatherer ancestors from the moment they achieved the ability to evaluate the outcomes of their innovative decisions within a single person's lifetime, often on a scale of months or days (if not hours or minutes), and sometimes even *ex ante*.

In consequence, species evolution explains the human history to the point of the cognitive revolution of the Upper Paleolithic but cannot explain any of the later developments. So why did they appear?

2.2 The Meaning of Life

Before we get there, it's beneficial to make a short detour. Let me ask you a quick warm-up question: what is the purpose of human life? What gives it meaning?

If you have a hard time answering this question, don't despair. You're not alone. There is no consensus in the scientific literature on this question and no sign that one will emerge in the foreseeable future—even though the problem has been tackled from a variety of angles, by philosophers, psychologists and neuroscientists alike. Also, ready-made dogmatic answers are preached by religions, aiming to dispel the confusion and focus our thoughts around one *certain* answer.

First, one may draw some ideas on the purpose of life from evolution theory. Darwinism, building on Charles Darwin's seminal theory (Darwin, 1859), suggests that the purpose of human life is merely to survive and multiply, thus carrying the human species to the next generation and allowing for the recombination of our genetic material. The drives to survive and multiply are built in our basic instincts that we share with other animals. An extension of Darwinism due to Richard Dawkins says that even the concept of instinct is perhaps already too complex; instead, the purpose of our lives is determined directly by our genes (Dawkins, 1976): our "selfish" genes hijack our bodies to recombine their information and forward themselves into the future. We are not agents, we are vehicles that transfer certain information—our genomes—across time, and recombine this information with other persons' information, potentially leading to evolutionary progress. We may subjectively perceive ourselves as serving noble goals and higher causes, but this perception is only an illusion, an adaptation helping us and our genes survive and multiply. Importantly, the Darwinian perspective works both at the individual and the aggregate scale. "Selfish replicators" such as people and their genes are naturally willing to cooperate with one another due to at least five mechanisms: kin selection, direct reciprocity, indirect reciprocity, network reciprocity and group selection (Nowak, 2006).

Second, a perspective specific to the economics orthodoxy views people as rational agents who consciously maximize their expected utility, derived from the consumption of various goods and the enjoyment of leisure. In extended setups, utility is sometimes also allowed to flow from other sources such as interpersonal contacts or the well-being of one's offspring. Experimental evidence, however, enumerates plentiful discrepancies between the expected utility maximization model and actual behaviors of human subjects in the lab (Kahneman et al., 1982; Ariely, 2008). We repeatedly disappoint the standards of rational decision making that economists would like us to maintain by showing inconsistent preferences (you may prefer A over B, B over C and C over A), failing to respect the axiom of independence from irrelevant alternatives (you may prefer A over B and C, but B over A if C is absent), or anchoring on the *status quo* (you may like a lottery that gives you 10$ with a 50% chance, but not one that gives you the 10$ first and then takes it back with a 50% chance). Nevertheless, there remains the possibility that the economists are right after all when it comes to the purpose of human life, only that they have not yet built a proper encompassing structure that would be consistent with all the available evidence.

Third, there is a range of religious answers. Many people believe that the meaning of human life is to praise and serve God, to redeem one's sins and go to heaven, to liberate oneself from earthly desires and achieve profound peace of mind, or to achieve some other forms of "transcendence". These answers are revealed in holy scripts as epiphanies and encapsulated in elaborate rituals. Inspiring as they are too many, they nevertheless fail to provide a convincing proof that their answer to the question of purpose of life is factually correct. Therefore nowadays the number of people challenging religious dogmas is mounting, in parallel with the volume of empirical observations on people's actions and reported motivations that are not aligned with the increasingly outdated religious teachings. (And also because in most of the world religion is not as brutally enforced anymore.) A notable atheist synthesis of these varying religious standpoints is that there may in fact exist an underlying objective, fact-driven ethical rule that people tend to adopt as their life purpose without conscious comprehension, such as "to maximize the well-being of all conscious creatures" (Harris, 2010). Religious scripts are written so as to (partially and imperfectly) embrace this rule and also to prepare room for establishing institutions able to control people by distilling their moral intuitions into religious dogmas.

Fourth, the humanist point of view emphasizes people's free will to set their objectives and define the purpose of their lives freely (Adler, 1946; Harari, 2017). Humanism rejects the false certainty of doctrinal answers and empowers people to make up their minds on this sensitive issue on their own. (Similar sentiments are also shared by existential psychology, Reker (2000).) This freedom and responsibility may make us agonize over this dilemma for days on end, but may also lead to practical insights on our relative strengths and weaknesses. Quite often, however, we would come up with a superficial answer that only magnifies the confusion on the original question—one that is equally hazy and unfalsifiable as religious answers are. For example, you may conclude that the purpose of your life is to seek happiness, find balance in life or become a better person. Noble and valuable, but do you really understand what this means to you and how you should act on these objectives? And are you sure that you won't reinterpret your conclusions 10 or 20 years along the road, faced with new knowledge and new life circumstances?

Fifth, a recent philosophical line of thought, backed by preliminary evidence from neuroscience, suggests that free will may be in fact only an (evolutionarily useful) illusion and our actions may be instead just a product of pre-determined causes and random events, neither of which we have control over (Harris, 2010, 2012; Harari, 2017). In such a framework, the humanist request to find one's personal purpose in life is ill-posed; one can still ask the question about people's *revealed* objectives nevertheless. This point of view suggests that our seemingly deliberate decisions may not be a product of the conscious self, but reflect the unconscious complex optimization mechanisms going on in our heads, shaped by our genes, culture and environment. Still, looking back at people's past choices provides useful positive information about us as a species: we don't consciously maximize any objective, but we act *as if* we maximized something. What is it?

The aforementioned five perspectives vary not only in their assessment of the purpose of human life, but also span the space left for our own conscious choice of objectives. The evolutionary perspective suggests that our objectives are hardwired in our minds and will be followed regardless of our opinion. The economic modeling view takes a similar stance but also assumes that we are aware and fully endorse our objectives. The religious perspective typically not only allows some degree of free will but also builds strong incentives to follow strict prescriptions for goals and actions. The humanist perspective, in contrast, is the only one that openly allows for independent deliberation of one's aims and is ready to accept varying answers. This perspective also stands out as the only one that is interested both in the positive side of the conundrum (what *is* the purpose of human life?) and its normative side (what *should* it be?).

The discussion on free will and purpose of human life is further complicated by the "hard problem of consciousness" (Chalmers, 1995), i.e., the question of how and why we happen to have subjective phenomenal experiences. What makes us conscious and able to make our decisions (freely or not)? How is consciousness embedded in our brain? Does it naturally emerge from its complexity or is there some additional component involved (Penrose, 1989; Tononi et al., 2016; Dennett, 2017)?

Furthermore, meaningful candidates for our ultimate life goals must avoid circularity. It is less than satisfying to say that the purpose of human life is to seek happiness, life satisfaction, fulfilment, self-realization, sense of purpose, authentic existence or the like. This just tautologically states that *the goal* is to successfully achieve *the goal*. Take, for example, happiness. Many things can make us happy, from the incidence of being born with a high base level of serotonin, through short-lived pleasures like having a good meal or seeing your home team win, to long-lasting life satisfaction which follows from, say, having a "meaningful" job and being "successful" at it, or being loved in one's family and respected in a broader community. Hence, happiness is a sum of more or less random external factors, and success in following *the goal*. Whatever it is.

In sum, science and philosophy have not quite figured out the purpose of human life yet, neither for the individual nor for the collective. There are several possible explanations for this incidence. First, a unique human objective (or a system of objectives) may exist but be unknown yet, perhaps too complex for us to learn it at this stage. Second, people's lives may have different purposes, and this variation may defy the attempts at finding a universal underlying objective. (For example, according to the socio-analytical theory (Hogan, 1983), although human nature is largely biologically determined, the survival of our species is predicated on individual differences, in particular with regard to life goals.) Third, purposes of people's lives may have evolved over the millennia, further complicating the endeavor. Fourth, for some or all of us there may even be no objective meaning of life at all, and collective outcomes may follow primarily from complex interactions in groups rather than goals pursued by individuals.

But what if, from the perspective of positive science, *it doesn't even matter* what the purpose of human life is? Perhaps a large array of answers to the fundamental question is observationally equivalent, that is, leads to exactly the same societal, economic and technological outcomes, making it impossible to reverse engineer the correct answer by studying human behavior? Let's investigate this possibility.

2.3 Instrumental Convergence

A surprising insight on the "what's this life for" dilemma can be drawn from the research on artificial intelligence. AI algorithms developed thus far tend to follow goals that are precisely specified in their code, such as "drive from A to B like an experienced, cautious human driver would drive" or "win the chess game". Humans and other biological species aren't as transparent—our brain software is certainly not *open source*—and therefore in our case we can only try to imperfectly reverse engineer our objectives by looking at actions and outcomes. Transparency and relative simplicity of the AI case makes it a perfect laboratory to observe that a wide range of optimizing algorithms, that chase very different goals, tend to share a surprisingly large number of common features. Now it only requires a small stretch of imagination to reason by indirect inference that humans and other biological beings should share these common features, too—as long as we also chase some generic goals, at least implicitly. Crucially, the content of the goals—as long as they're not *unusual in a very specific sense*—doesn't matter that much: in particular, goals may vary across individuals, change over time and be subject to statistical noise.

The *instrumental convergence thesis* (Fig. 2.1), formulated and discussed at length by Omohundro (2008) and Bostrom (2012), applies to AI algorithms with almost any final goal, and particularly so to complex goal systems that don't admit a global optimum to be reached in finite time. (All known counterexamples to instrumental convergence are based on carefully crafted, highly stylized, simple goals.) It states that regardless of the final goal, the algorithm will also pursue the following instrumental goals: (i) self-preservation (including goal-content integrity), (ii) technological perfection/efficiency, (iii) cognitive enhancement / creativity, and (iv) resource acquisition.

In my perception, Steven Omohundro and Nick Bostrom may have actually discovered a deeper truth about the world than the AI context of their thesis would

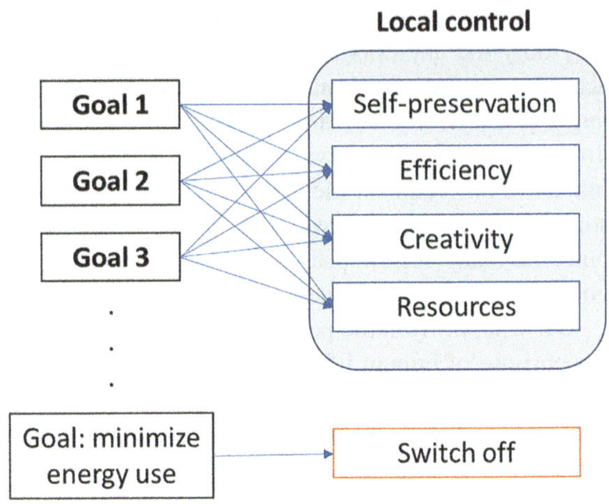

Fig. 2.1 Instrumental convergence. *Note* instrumental convergence works for *almost any* final goal but not all possible goals. Some *simple* goals can be achieved fully without requiring any instrumental goals. For example, minimum energy use (zero) is achieved by immediately switching off

suggest. It appears to me that not just AI, but also the biological world at large, from bacteria to humans, displays instrumental convergence. Naturally, besides AI this finding is most interesting when applied to humans, and not because we are unique in sharing it with AI, but rather because we have the largest capacity to pursue the four instrumental goals effectively. We owe this privilege to the well-developed frontal cortex of our brains.

Self-preservation is the most basic auxiliary goal. It works regardless of whether you are a bacterium, an animal, a human being or a computer algorithm: simply, whatever your purpose is, you will no longer pursue it when you're dead. Moreover, if your goal is a long-standing and open-ended one that is never fully completed, it also makes a lot of sense to multiply. If your offspring aligns with your goal, you will in this way both increase the combined optimization power employed in pursuing the goal at present, and ensure that your goal will remain in place after your death. (Alignment of goals between parents and offspring is guaranteed in the case of an algorithm that copies its code. In the case of sexually reproducing species, however, goals transfer across generations quite imperfectly.)

Self-preservation includes also the will to *preserve goal content*: a good optimizer is faithful to its objective because any change in the goal compromises the original goal. For example, if your life is ultimately dedicated to producing guitars, you would like to survive as long as possible (in order to produce more guitars), you would like to spread the goal of guitar production as wide as possible (also by having offspring and instilling guitar production in them), and you will be never convinced to switch to producing violins (because that would be harmful to what you care about, that is, producing guitars).

Taking the Darwinist approach, one may view self-preservation and multiplication as the unique goal of our (and other) species. It is instructive to observe, however, that this goal will be followed also if the true final goal is completely different—for example to produce guitars. Self-preservation emerges instrumentally from a wide variety of final goals.

The second instrumental goal is *technological perfection*, also referred to as efficiency. Whatever the final goal is, as long as the available resources are limited, it is best achieved if the resources are used efficiently. "Resources" in this context may refer to not only raw materials, machines and other tools but also energy and time. For example, if your life is ultimately dedicated to producing guitars, you would like to organize the technological process of guitar building efficiently so that you would waste as little of your materials and energy as possible, and work as focused and quick as possible so that you can produce as many guitars as you can under the binding constraints.

The third auxiliary goal is *cognitive enhancement*, also referred to as creativity. Whatever the goal is, and especially if it is long-standing and open-ended, it pays to devote part of the time and resources to improving one's performance in pursuing the goal. The short-run opportunity costs will be then, in expectation, covered by the long-term performance gains. You may wish, for example, to learn new skills related to the goal or perform experiments which would verify if certain novel ideas are helpful in improving your performance. Turning to the guitar example again: if your life is ultimately dedicated to producing guitars, you would be clearly interested in improving the technological process of guitar building. You'd be happy to invent and adopt new ways of producing guitars that would be more efficient than the ones you used so far.

Finally, there is the instrumental goal of *resource acquisition*. Whatever the goal is, it pays to throw more resources at it. For example, if you produce guitars, you would easily produce more of them when provided with a larger supply of intermediate resources such as wood, plastic, metal frets and strings. And you'd be happy to establish subsidiary companies which would transform raw materials into these intermediate resources for you, allowing you to produce even more.

Importantly, this auxiliary goal follows even from goals that impose explicit resource restrictions. For example, if your goal were to produce as many guitars as possible *from a single box of wood*, you would logically still prefer to use more wood, for experimentation and learning, so that you would in the end—by trial and error—learn to produce guitars from a single box of wood as efficiently as possible. Not to mention that you'd soon discover a thrilling new way around the resource restriction: to replace scarce wood in the guitar design with other, less limited resources, such as metal or plastics.

2.4 Like It or Not, You're in It

The bottom line of this discussion is that *whether you like it or not, you're in it*: whatever the ultimate objectives of your life are, you must also pursue the four instrumental goals. We're all deep in this, really, and at all levels: both individually, in groups, as nations and as a species.

In the remainder of this book—for the sake of a convenient shorthand—by *local control maximization* I will call the process of simultaneously pursuing the four instrumental goals. Maximizing local control means (i) putting yourself on safe ground, (ii) exploring the space of possible actions in order to reflect your preferences as closely as possible and satisfy as many needs as possible with the resources available in your environment, (iii) investigating the frontier of your technological capabilities in search for better ways of satisfying your preferences and needs in the future (iv) and actively seeking additional resources.

My thesis is that local control maximization applies not only to AI algorithms (Omohundro, 2008; Bostrom, 2012) but also broadly to biological life and specifically to humans. When applied to the *homo sapiens*, local control maximization explains the broad historical trends in terms of human actions and declared incentives, observed across all human history. Whatever the true purpose of human life is, as long as it's consistent with the conditions for instrumental convergence, it is conducive to technological, economic and societal development.

The umbrella word "control" used in naming the concept signifies that the underlying objectives may include a variety of variables and motives; what unifies them is our quest for control over them. (Unsurprisingly, psychological literature finds that autonomy—the freedom of choice and control—is an important determinant of people's life satisfaction and happiness, Nussbaum (2013), Inglehart (2018).) The word "local", in turn, is meant to emphasize that our information sets and planning horizons are limited, both in space, so that we take only part of our surroundings (natural environment, other people) into account, and in time: we're myopic and can't properly handle the uncertainty with respect to future developments. (This also explains many unintended consequences of human actions such as mass extinctions, civilizational diseases and climate change. We just hardly ever think about the whole mankind or the infinite time horizon.)

Instrumental convergence may emerge both in conscious and unconscious processes, running both in the biological and the silicon-based substrate. Postulating local control maximization as a universal driving force of development is therefore unrelated to the resolution of the "hard problem of consciousness", that is, how and why sentient organisms have phenomenal experiences (Chalmers, 1995). Similarly, my thesis also circumvents such deep questions as, does a conscious mind arise out of any sufficiently complex data processing unit, or does it require special architecture? (So far, the most promising hypothesis is that consciousness arises in systems able to harbor a sufficient degree of *integrated information*, Tononi et al. (2016).) Does consciousness allow one to modify one's goals? These problems are extremely interesting in their own right, and will have a bite in the future with advanced AI, but are probably irrelevant for explaining the human history so far and our immediate future.

It is also instructive to juxtapose local control maximization and intelligence. The latter can be defined, among other definitions, as *efficient cross-domain optimization power*, the ability to hit narrow targets in broad search spaces (Yudkowsky, 2013). Hence, intelligence measures how powerful a given decision-maker (whether human, animal or computer algorithm) is in performing its local control maximization. Intelligence doesn't affect *what* we maximize; it determines *how good* we are at it. In fact, Nick Bostrom's orthogonality thesis states that any level of intelligence can be applied to essentially any final goal (Bostrom, 2012).

The beauty of the local control maximization process lies in its emergent character. It is a generic outcome of instrumental convergence. Individual decision-makers don't have to consciously follow the objective of local control maximization (and in the case of humans, people often actively oppose it, declaring instead the pursuit of happiness, harmonious family life etc., or recalling religious or spiritual motivations). For instrumental convergence to emerge at the macro scale, in a network of decision-makers, some of them may even have no ultimate goal (no objective *meaning of life*) whatsoever or their goal system may be inconsistent and internally contradictory. (You may wonder if animals also pursue "higher" instrumental goals like technological perfection or cognitive enhancement. They do. They're just bad at it.)

There is a nice analogy between local control and the normal distribution, or bell curve. In a wide range of natural, social and economic phenomena, individual units tend to be located along a very parsimoniously described bell curve. Air temperatures specific for location and season are normally distributed, people's height in a population is normally distributed, crop yields are approximately normally distributed, and so are firm sizes when measured in logarithms. But this is not because nature and society have a special taste for bell curves. Rather, what generates this result is the Central Limit Theorem (CLT): if the measured units arise as a sum of a large number of independent components, then under very weak and generic assumptions their distribution will be normal. Just like CLT, the instrumental convergence thesis also has the power to morph general assumptions into specific conclusions. It says that under very weak and generic assumptions on the set of true goals of an optimizing entity, it will also emergently chase very parsimoniously specified local control.

2.5 Local Control and the Hierarchy of Needs

Local control maximization is shaping the development of human civilization by allowing us to satisfy our needs. The evolution of these needs over the millennia of human history tells us a lot about the workings of instrumental convergence in *homo sapiens*. Our modern lives are very different from our ancestral environment of a hunter–gatherer tribe in prehistoric Ethiopia precisely because the gradual increase in the extent of human control over the environment achieved across the five eras of economic and technological development between 70 000 BP and today allowed us to reach out to ever more sophisticated, advanced needs. This progress can be intuitively illustrated with Abraham Maslow's hierarchy of needs (Maslow, 1954).

Hunter–gatherers can only satisfy their most fundamental needs. They spend a large chunk of their time searching for food, a *physiological* need. Furthermore, as their rudimentary equipment, which they carry with themselves, provides them with some warmth and shelter, and they live in extended families as well as organize into bands and tribes, we may say that some of their basic *safety* and *social belonging* needs are also satisfied.

Compared to hunter–gatherers, sedentary subsistence farmers can satisfy somewhat more *safety* needs, thanks to their permanent dwellings, food storage, elimination of many threats (e.g. predators, hostile tribes) from the ecosystem, and the accumulation of durable goods. Following the increases in population density and growth of cities and states, *social belonging* needs can be potentially better satisfied, too.

The scientific revolution allowed for a wider spread of education and opened new opportunities for human creativity and inquisitiveness, allowing the (still relatively narrow group of) individuals who were not bound to food production to pursue the needs of *esteem* and *self-actualization*.

The industrial revolution was a real game-changer in this regard, opening up a wide range of new possibilities for satisfying human needs. Sustained growth in per capita output allowed a wide group of people to benefit from increased standards of living and financial security, for the first time in history. Developments in medicine dramatically improved human health and increased our longevity. All this can be classified as satisfaction of additional *safety* needs. The industrial economy also greatly widened the group of people who could develop their specific skills and gain recognition for their work, thus satisfying *esteem* and *self-actualization* needs.

Compared to the industrial era, the digital era offers additional avenues for satisfying *esteem* and *self-actualization* needs. The digital sphere offers high rewards for creativity, innovativeness and unique skills. The Internet democratizes access to platforms of self-presentation and facilitates recognition of one's work. Finally, whether we like it or not, gradual automation of routine jobs pushes us away from the most boring, repetitive occupations towards more skilled and creative work.

Our species' systematic advancement up Maslow's hierarchy of needs (Maslow, 1954) is an indication of success in pursuing Omohundro's and Bostrom's instrumental goals. Satisfying low-level needs (physiological, safety and social belonging) follows the goal of self-preservation. The drive for resource acquisition helps satisfy both the low-level need of safety and the high-level need of esteem. (Note that the resources we accumulate may vary from an inventory of consumption goods and financial wealth to social contacts, political power, and useful information.) The drive toward creativity (cognitive enhancement) and efficiency (technological perfection) is, in turn, linked to the high-level needs of esteem and self-actualization. Both are achieved by engaging in activities such as learning, experimentation, research and communicating with others in a social network.

Comparing the emergent instrumental goals to Maslow's categories of human needs we observe, however, that some needs instead of serving the goals, may act as substitutes for our local control when it is unattainable, for example when there is uncertainty which we cannot eliminate. This includes the *social belonging* needs of bonding and emotional support, served, e.g., by maintaining close kinship ties (Growiec and Growiec, 2014). Accordingly, holding religious or superstitious beliefs is our paradoxical way to get around fundamental uncertainties and satisfy some of our *safety* needs without actually increasing local control.

What remains unresolved is the mystery of the *actual final goal*. Advancing up the Maslow's hierarchy of needs surely feels great, but in itself doesn't necessarily bring us closer to fulfilling the true meaning of human life, whatever it is; it only marks our increased ability to pursue our *instrumental* goals. A question remains then, does progress in pursuing instrumental goals imply progress in pursuing the final goal?

Logically, the answer should be positive: the purpose of instrumental goals is, by definition, to serve the final goal. If chasing them contradicted our final goal, we wouldn't be doing that. (Unless in the course of development, our original final goal has been replaced by an instrumental goal—an event that Eliezer Yudkowsky calls a "subgoal stomp".) Hence, my hypothesis in this regard—that still awaits empirical verification—is that the technological and economic development of human civilization has been not only facilitating satisfaction of our instrumental needs, first the basic ones and then progressively the higher-order ones, but has also been helpful in pursuing our true final goal, both individually and collectively as a species. The hypothesis implies that local control maximization, reflected by global population growth, accumulation of wealth and progress in science, was only a necessary milestone on our way towards realizing our "true human purpose". If the hypothesis is true, then the future behavior of the humankind will converge towards our final goal, gradually revealing it from the fogs of local control.

2.6 Origins of the Human Civilization

But if instrumental convergence is a generic outcome, why are humans so much more successful in pursuing local control maximization than other living beings? How have we come to be the only species in Earth's history to dominate the entire planet, modify most of the world's ecosystems and drive other species to mass extinction (Kolbert, 2014)? After all, we are clearly not unique in our wish to satisfy our various needs, avoid risks, pass our genes to the next generation, act efficiently and accumulate resources. Chimpanzees and other great apes show us that we are also not unique in looking for creative solutions improving our performance and expanding the set of available choices. So what distinguishes us? What is the key gift that empowered us to escape the fate of other animal species: eternal struggle for survival and reproduction that consumes almost all time and energy?

Apparently, the key anatomic feature that enabled this unprecedented success is the frontal cortex of our brains. Its development allowed us not only to have relatively smart ideas but also—crucially—to join forces with others to improve upon them. Once our cortical capacity has surpassed a critical threshold, the *homo sapiens* acquired sufficiently advanced communication skills and theory of mind to be able to document and share stories, gossip, tales, abstract ideas, and ultimately knowledge and technology (Dunbar, 1998; Tomasello, 2014; Harari, 2014). The balance between information acquired and lost in a unit of time turned from negative to positive. Our ability to use language to share not only factual but also fictitious information, and to maintain social ties, led our abilities to a unique level of versatility and adaptivity, inaccessible to any other species. It allowed us to organize ourselves into bands, tribes, local communities, and ultimately societies and nations—and cooperate *flexibly* at varying scales, from local to global. It was a key developmental milestone that enabled our species to spread, conquer, and eventually rule the whole world.

Bigger frontal cortex of our brains allowed our species to develop complex, intricate social networks, whose coherence is ascertained by using language. The human brain can naturally accommodate social contacts up to ca. 150 acquaintances (Dunbar, 1992, 1993) which is really a lot when compared to other species: the runner-up is the chimpanzee who can do 50—and has a rather tight time budget because it still relies on grooming instead of talking to maintain its social ties (Dunbar, 1998). The "Dunbar's number" of 150 not only determined the size of hunter–gatherer tribes in the past, but works well even today: for example, it sets the maximum size of a well-functioning organization before it requires a hierarchical structure.

The social character of our species manifests itself both in our actions and their perceived motivations. Social belonging is one of our key needs, and being loved and respected are important for our happiness. We intuitively understand our current social status and adjust our actions to maintain or improve it. Our self-esteem acts as a sociometer (Leary et al., 1995) measuring the extent to which we are included versus excluded by others. A biologically determined mechanism then associates social exclusion with a threat to one's survival. According to the psychological Terror Management Theory (Solomon et al., 1991) our self-esteem acts also as a cultural buffer against the fear of mortality. Our brains are capable of acting both selfishly and toward the good of larger groups to which we belong. And being able to theorize about the minds of others, for example our children, we are also able to consider the future of the world after we die.

When the brains of our ancestors first crossed the threshold of *cumulative knowledge accumulation*, the net balance of non-genetic information gained and lost by the humankind in the time span of a generation turned from negative to positive, taking the human species from stagnation to accelerating growth. That moment, about 70 000 years ago, was the cognitive revolution. Tight constraints limiting our capability for exercising local control over our environment were finally loosened and we were no longer at the mercy of species evolution—positive changes could finally be implemented and sustained without being imprinted in our genome. The genie was out of the bottle. The human civilization has begun.

References

Adler, A. (1946). *Understanding human nature*. Greenberg.

Ariely, D. (2008). *Predictably irrational: The hidden forces that shape our decisions*. Harper Collins.

Ashraf, Q., & Galor, O. (2013). The "Out of Africa" hypothesis, human genetic diversity, and comparative economic development. *American Economic Review, 103*, 1–46.

Bostrom, N. (2012). The superintelligent will: Motivation and instrumental rationality in advanced artificial agents. *Minds and Machines,22*, 71–85.
Chalmers, D. (1995). Facing up to the problem of consciousness. *Journal of Consciousness Studies,2*, 200–219.
Darwin, C. (1859). *The origin of species*. John Murray.
Dawkins, R. (1976). *The selfish gene*. Oxford University Press.
Dennett, D. (2017). *From bacteria to Bach and back: The evolution of minds*. Norton & Company.
Dunbar, R. I. M. (1992). Neocortex size as a constraint on group size in primates. *Journal of Human Evolution,22*, 469–493.
Dunbar, R. I. M. (1993). Coevolution of neocortex size, group size, and language in humans. *Behavioral and Brain Sciences,16*, 681–735.
Dunbar, R. I. M. (1998). *Grooming, gossip, and the evolution of language*. Harvard University Press.
Growiec, K., & Growiec, J. (2014). Trusting only whom you know, knowing only whom you trust: The joint impact of social capital and trust on happiness in CEE countries. *Journal of Happiness Studies,15*, 1015–1040.
Harari, Y. N. (2014). *Sapiens: A brief history of humankind*. Vintage.
Harari, Y. N. (2017). *Homo deus: A brief history of tomorrow*. Vintage.
Harris, S. (2010). *The moral landscape: How science can determine human values*. Free Press.
Harris, S. (2012). *Free will*. Free Press.
Hogan, R. (1983). A socioanalytic theory of personality. In M. Page & R. A. Dienstbier (Eds.), *Nebraska symposium on motivation personality—current theory and research* (pp. 55–89). University of Nebraska Press.
Inglehart, R. F. (2018). *Cultural evolution: People's motivations are changing, and reshaping the world*. Cambridge University Press.
Kahneman, D., Slovic, P., & Tversky, A. (1982). *Judgment under Uncertainty: Heuristics and Biases*. Cambridge University Press.
Kolbert, E. (2014). *The sixth extinction: An unnatural history*. Henry Holt and Company.
Leary, M. R., Tambor, E. S., Terdal, S. K., & Downs, D. L. (1995). Self-esteem as an interpersonal monitor: The sociometer hypothesis. *Journal of Personality and Social Psychology,68*, 518–530.
Maslow, A. (1954). *Motivation and personality*. Harper & Brothers.
Nowak, M. A. (2006). Five rules for the evolution of cooperation. *Science,314*, 1560–1563.
Nussbaum, M. (2013). *Creating capabilities: The human development approach*. Belknap Press, An Imprint of Harvard University Press.
Omohundro, S. M. (2008). The basic AI drives. In P. Wang, B. Goertzel, & S. Franklin (Eds.), *Proceedings of the First AGI Conference, Volume 171, Frontiers in Artificial Intelligence and Applications*. IOS Press.
Penrose, R. (1989). *The emperor's new mind: Concerning computers, minds and the laws of physics*. Oxford University Press.
Reker, G. T. (2000). Theoretical perspective, dimensions, and measurement of existential meaning. In G. T. Reker & K. Chamberlain (Eds.), *Exploring existential meaning* (pp. 39–55). Sage.
Solomon, S., Greenberg, J., & Pyszczynski, T. (1991). A terror management theory of social behavior: The psychological foundations of self-esteem and cultural worldviews. *Advances in Experimental Social Psychology,24*, 93–159.
Tomasello, M. (2014). *A natural history of human thinking*. Harvard University Press.
Tononi, G., Boly, M., Massimini, M., & Koch, C. (2016). Integrated information theory: from consciousness to its physical substrate. *Nature Reviews Neuroscience,17*, 450–461.
Yudkowsky, E. (2013). *Intelligence explosion microeconomics* (Technical report 2013-1). Machine Intelligence Research Institute.

Chapter 3
The Force Unleashed

Be fruitful and multiply, and replenish the earth, and subdue it; and have dominion over the fish of the sea, and over the fowl of the air, and over every living thing that moveth upon the earth.

Genesis 1:28

3.1 Filters Behind and Before Us

By giving our species brains with an upgraded frontal cortex evolution helped us escape the eternal struggle for survival and reproduction—the fate of other species—and empowered us to pursue local control maximization across a wide spectrum of choices. But was the rise of humankind inevitable? If instrumental convergence in data processing units "by nature" leads to local control maximization, and local control maximization "by nature" leads to economic and technological development, then the answer should be positive, right?

Well, not exactly—and not only because the humankind could have been at some point wiped out by a massive asteroid impact, a global volcanic event or a global nuclear war. The key point here is that while being an emergent, powerful driving force of development, local control maximization doesn't guarantee *unbounded* growth. There is always a possibility that development permanently stagnates or reverses. To reach the current level of wealth and technological sophistication, the human civilization had to clear a few critical development thresholds, or "filters" (Hanson, 1998). Neither of them was easy to pass.

The first filter after the emergence of *homo sapiens* was the cognitive revolution. For our species to win the competition against all other animals and dominate all Earth's ecosystems, the frontal cortex capacity of our brains must have surpassed the critical threshold of net positive knowledge accumulation by the interconnected human collective. The volume of new valuable information acquired by each

J. Growiec, *Accelerating Economic Growth*, Frontiers in Economic History,
https://doi.org/10.1007/978-3-031-07195-9_3

consecutive generation and passed to the next one must have exceeded the volume of information lost through forgetting and death. Clearing this hurdle was *ex ante* extremely unlikely, as evidenced by the fact that no other surviving species managed it.

The next filter was to start the agricultural era. To achieve this, our ancestors we must have succeeded in domesticating some key crops and animals. This too was rather unlikely *ex ante*, and failure was possible on both sides of the equation, the people and the domesticable species. For example, Aboriginal Australians—anatomically identical to Asians, Europeans or Native Americans, and just as capable of initiating agriculture—happened to live in a relatively unwelcoming environment where no crops were available for domestication, and therefore they never managed an agricultural revolution of their own making (Diamond, 1997). If the humankind had failed at this filter globally, development could have stopped at the hunter–gatherer level for hundreds of thousands of years—perhaps until a global catastrophe happens or a new species evolves and supersedes humans. Apparently the good fate of humans was in this regard decided by the existence of wild wheat, rice and corn (and other founding crops of agricultural revolutions).

Down the road, further filters could be placed before the dawn of each consecutive era: neither the scientific, industrial nor the digital revolution was *ex ante* bound to happen. For example, if the worldwide area of arable land were smaller, the humankind could have potentially stagnated in the permanent Middle Ages, with a labor-intensive agricultural technology and a highly polarized society of a narrow landed elite and a mass of landless peasants. (Some say that the same outcome was also likely, had Europe been spared from the fourteenth-century Black Death pandemic.) Similarly, if the global volume of energy resource deposits such as oil or coal was much smaller, we could have used it all up before inventing, e.g., electricity, nuclear power or computers. Without these technologies and without the fuel for further industrial production, there could have been no other way but back to an agricultural economy.

Filters that end up in stagnation are bad but even worse are the ones that lead to a decline or outright doom. For example, the isolated population of Easter Island managed to build a sophisticated civilization with artifacts like the *moai* that stun the observers to this day, but in the process it depleted most of its resources. After the island had suffered complete deforestation and profound soil erosion, its population declined massively, perhaps fivefold or more (from about 15 000 to 2 000–3 000). The island was forced to abandon the path of growth and development, and its civilization stagnated until the day European explorers first charted the island in 1722 (de la Croix and Dottori, 2008). It can be imagined that the same story—only magnified by a factor of a million—could repeat for the entire Earth as well if we, for example, deplete our exhaustible resources and cause irreversible climate change. Is this a realistic scenario? Hard to say: on the one hand, we have much more advanced technology now, much more substitution possibilities (e.g. nuclear, solar, wind and hydropower instead of power from fossil fuels), and much greater research potential which may help us in need. On the other hand, in a world filled with nuclear weapons conflicts over scarce resources always pose a major threat, and global warming by a few degrees centigrade during the current century—a quite possible scenario when accelerating growth is coupled with countries' reluctance to limit greenhouse gas emissions—can be really earthshaking.

Predicting future filters is of course a risky business. For one thing, though, economics teaches us that scarcity of a factor motivates our ingenuity and creativity in overcoming or circumventing it. Scarcity of food possible to obtain from a given ecosystem by hunting and gathering focused directed efforts to transform the ecosystem and eventually allowed our ancestors to domesticate plants. Scarcity of land in the Old World in the late Middle Ages focused people to engage in geographic exploration, develop international trade and invent new farming technologies able to get around land scarcity. Scarcity of labor in the early industrial era, in turn, allowed to develop machines that substituted for it. The mechanism is simple: as the scarce factor is becoming more and more expensive, it becomes more and more profitable to shift an increasing fraction of time and resources from the "business as usual" scenario to research and experimentation on new goods and processes that would serve the same purposes but replace the scarce factor with something cheaper and more abundant. In the economy like in nature: scarcity provides vital incentives for adaptation to a changing environment. For example, the oil crisis of the 1970s generated a huge spike in oil prices, motivating motor companies to build less fuel-consuming vehicles and governments to invest more in alternative sources of energy.

Therefore perhaps not all of what we may now perceive as possible future filters are genuine filters. We may, for example, learn to live without fossil fuels and still produce as much energy as now, or maybe even a million times more. Human ingenuity could perhaps one day produce clever ways to survive and thrive in a heated up world—or reverse global warming altogether. This we will be able to say for sure only after we've passed the potential filter, though.

The fundamental question remains: Are there any *genuine* filters still ahead of us? If so, what are they? Furthermore, since our imagination already reaches the stars, is there still a major filter between the human civilization of today and the prospects of colonizing other planets, stars and galaxies? Does a Great Filter exist, and if so, is it behind us or before us?

Robin Hanson first formulated the concept of the Great Filter while thinking about the Fermi Paradox (Hanson, 1998): how come don't we see any signs of extraterrestrial intelligence? The universe is so huge and varied; why does it seem so dead? About 13.8 billion years have elapsed since the beginning of the universe, but since our planet is only 4.5 billion years old and our species has colonized it and built a modern technological civilization in just 70 000 years, other civilizations may have had billions of years of a headstart and by now may be technologically more capable by many orders of magnitude. So why don't we see any advanced civilization out there?

Well, as Robin Hanson argues, if there exists an almost unpassable filter—the Great Filter—or a sequence of such filters, before a civilization can colonize the universe, then only a tiny fraction of them would be able to clear it. And if they fail to clear it, they would either stagnate or decline and never form a civilization that we could be able to observe from afar. The implication for the humankind—which has thus far followed a path of accelerating growth—is that the easier it was for life to evolve to our stage, the smaller our chances probably are for further development and even survival in the future.

Oh, and if you wanted a clear answer to whether we've passed the Great Filter yet, here it is. We don't know.

3.2 Evolution, Here's a Surprise!

For millions of years life on Earth evolved in a process of continuous adaptation and conflict over the ecosystem's resources. Species developed from simple to complex and gradually acquired sophisticated adaptations allowing them to win an edge over certain other species and capture more resources. Species that were driven out of their ecological niches had to become extinct, and new species arose and took their place. It was a reality of constant war, but a war that had many fronts and no side was able to win on all of them. All species sought to maximize local control over their surroundings but were heavily constrained by the actions of other species which had the same (instrumental) goals. None of them could make a real difference because they were not able to pass the filter of net positive knowledge accumulation. With insufficient brainpower and communication skills, there was no other vehicle than the genome to pass valuable innovations across generations and enable other individuals to improve upon them. Therefore the skill set of a species was generally fixed over time, and the only force able to expand it was evolution and natural selection of individuals that were lucky enough to receive a genetic mutation that better adapted them to the environment.

When humans entered the stage and passed the filter of net positive knowledge accumulation, our drive to maximize local control was finally set free. With an enhanced theory of mind, language, and the resulting ability to create, document and communicate stories, gossip, tales, legends and abstract ideas (Dunbar, 1998; Harari, 2014), humans have become finally able to join their forces in information processing. Valuable innovations were passed across generations not just in the genome, but also through speech and demonstration, recursively augmenting our species' *immortal collective brain*. Next generations didn't have to start from scratch because development was a recursive, step-by-step process. In a world of highly isolated, small hunter–gatherer bands and tribes, this still meant a timeframe of hundreds of years to develop even the simplest of tools, and still a lot of valuable knowledge was lost in the process. Nevertheless, this pace of technological progress was already orders of magnitude faster than the pace of evolution—for which the timescale is measured in *hundreds of thousands* of years. Finally, by systematically applying the incremental changes, humans learned how to gain an edge over further and further species, spread and conquer additional resources. As a result, people multiplied, societies grew and became more complex, and species after species were brought to extinction (Kolbert, 2014). All of that happened without any need for alteration of the human body. Modern humans are anatomically (almost) identical to the tribes which lived in the

Ethiopian savanna 70 000 years ago. The *homo sapiens* is the first species on the face of Earth that doesn't have to hinge on evolution to achieve lasting progress. Once we've undergone the cognitive revolution, local control maximization serves us even better.

In other words, human local control maximization is an emergent sub-routine of the grand process of species evolution, and the only one so far which got out of evolution's hand. Evolution meticulously crafted a genie and put it in a bottle. Humans released it. And the reasons why we managed to do it are the following: (i) sheer pace of local control maximization—faster by orders of magnitude than the pace of evolution, (ii) ability for recursive self-improvement, a.k.a. our collective ability to learn from past mistakes and (iii) greed for resources, a.k.a. "I want it all and I want it now".

3.3 Technology and Human Control

While seeking to maximize our local control, we learned to build devices which augment our limited innate abilities. We carved sticks to reach further than our bare hands could reach, we built bows and arrows to hunt from a safe distance, and we documented facts and ideas on stone and paper so that we would not have to remember them later. Later on, we went on to build bicycles and cars so that we could move faster, and digital computers so that we would not have to invert that bloody four-by-four matrix by hand. As these devices got more and more sophisticated, we outsourced more and more of our actions. Sometimes these developments backfired on us, when we hurt ourselves with the stick, crashed the car into a tree, or realized that our job as a janitor has just been replaced with a surveillance camera. But generally, these innovations increased the span of human control.

The improvements may have affected the extent of human control either through augmenting our physical capabilities or brainpower. The first group of innovations allowed us to perform physical actions faster, stronger and more accurately, whereas the latter group helped us process information with greater pace, accuracy and depth. Within the first group, tools like a lever or a crane empowered us to lift heavy objects, and bicycles, cars and jet airplanes allowed us to travel at high speeds. In the second group, books and letters augmented our memory, thus allowing for more reliable storage of information, whereas pencil and paper empowered us to carry out deeper calculations.

New technologies extend human control by overcoming our limitations so that we can perform tasks that we couldn't do before. Some of them are relatively minor incremental developments, like upgrading a bow into a crossbow or a transistor screen into an LCD screen. Some of them, in contrast, like the five aforementioned major technological revolutions, have the potential to change the ecosystem, society and economy thoroughly and irreversibly. For example, the agricultural revolution allowed for the first time for productive transformations of the ecosystem, with a very human-centered understanding of "productivity"—namely measured in calories, to be consumed by people, produced per unit of land (Diamond, 1997). Side effects were of course abound, affecting nutrition, landscape and biodiversity alike, but hey, who cares about that stuff under the risk of starvation? After your ancestors had adopted farming and human populations had increased, you couldn't just go back to hunting and gathering; you'd die.

The last observation exemplifies the fact that, no matter how brilliant our ideas may be, our control maximization process is always really *local*: myopic and localized in space. Long-run, global side effects are nearly always beyond the horizon of events included in decision making. Even in the case of major political decisions deliberated in the present day by large groups of highly educated and well-meaning professionals, their combined (and technologically augmented) brainpower cannot handle the overall complexity of potential impact their decision can have on the world as a whole. Moreover, many people's actions involve experimentation by trial and error, not to mention the actions driven by lower motives.

It should be emphasized that technological advances are generally *irreversible*. I'm not aware of any case where useful technology has been consciously abandoned due to its unwanted side effects (where by "abandoning" I don't just mean replacing with a close substitute, as it was the case, e.g., with banning CFC freons in refrigerators or lead in gasoline). There have only been incidences when technological advances were reversed unwillingly and at a cost. For example, ancient Roman and Greek accomplishments in science and engineering were largely abandoned after the fall of the Roman empire in fifth century CE. At the same time, however, Europe witnessed a major population decline, driven both by violent "Barbarian" conquests and natural causes such as plagues. It was by no means a conscious decision not to use these technologies any more.

Another point is that human local control can be exercised either over the natural environment or other humans. Over the course of time we have gradually learned to control the natural environment. Concurrently, however, we have also increased the number of people with whom we regularly interact. Networks of friends and acquaintances have grown over time and become more intricate, eventually forming a "small world" global social network (Barabási and Albert, 1999). The latter fact mirrors both our huge reproductive success, easier satisfaction of other needs which freed our time and resources to engage in more social interactions, and recently the tremendous developments in communication technologies, allowing for instant information transmission and audiovisual contact with anyone across the globe. The consequence is that over time, and particularly in the digital era, the human local control maximization process is increasingly applied to interactions with other people and man-made machines rather than the natural environment.

One of the implications of this phenomenon is the democratization of access to information creation and sharing. Instead of relying on traditional ways of social status and hierarchy building, many of us now take matters into own hands, managing extensive digital social networks and creating and sharing original media content on platforms such as YouTube, Facebook, Twitter, Instagram, Snapchat etc. The average quality of these products notwithstanding, we are witnessing an unprecedented explosion of human creativity. Increased information creation and sharing bring measurable increases in aggregate human control: more useful data are created, collected and shared, more digital goods are consumed, and diffusion of useful technological knowledge is faster. And while some may dismiss many of those actions as economically unjustified (no direct monetary compensation), or view them as compulsive behaviors amidst an epidemic of narcissism (Twenge and Campbell, 2009), local control maximization is a mechanism that rationalizes these actions quite well.

3.4 Will AI Become a Filter?

Tools and machines always allowed humankind to increase the span of our control over the environment. Unfortunately, advanced artificial intelligence may put this reassuring regularity in question. Why would it?—one may ask—If AI is just a set of optimization algorithms created by human programmers, then it can only pursue goals set by human programmers. It is then just one more external device augmenting our abilities. So far this has always been the case; alas, it may change in the future.

There is an emerging general trend that as our technology becomes more and more complex, we agree to sacrifice some of our control at lower levels in exchange for more control at higher levels. For example, many of us no longer wish to know how various devices work as long as we know how to operate them and find the results useful. Who cares about microwave physics when the appliance is just so easy to use: you click here, wait a minute and your food is hot. Information technologies, in particular, are now beyond comprehension of *virtually anyone* in the world. Yet, click, and you watch an interesting movie or connect with a distant colleague who also has just logged on to Facebook. That's very empowering.

An important class of human capability augmentations is the ones that feature their own optimization algorithms. For millennia the only representatives of this class were domesticated animals. Cattle, pigs, sheep, horses, as well as our pets such as cats and dogs, are descendants of wild beasts that have evolved as local control maximizers, but we have learned to hijack their capabilities and channel them toward our own good. As the optimization power of animal brains is limited and constant over time, our superior position over them is unquestionable.

Things look different, though, with AI algorithms which are also goal maximizers and thus—like biological species—are subject to instrumental convergence (Bostrom, 2012). This is because unlike the animal species, their algorithmic complexity and computing power of the hardware on which they are run has been strongly increasing since the 1980s. A systematic trend has also emerged that AI increasingly makes decisions on our behalf and we are less and less able to understand the premises for these decisions—which may regard not only issues such as spending the next few hours of your free time (beware of the Internet binge) but also choosing the politicians you'd like to vote for (beware of fake news campaigns), and sometimes even your life partner (beware generally). As Arthur C. Clarke puts it, "any sufficiently advanced technology is indistinguishable from magic". Our digital technologies are clearly on track to become such in the near future.

People often underestimate how much progress has been obtained in the field of AI. Does anyone remember 1990s web search engines like AltaVista? It was nearly impossible to find the desired results. Now it's exactly the opposite: if ten top finds on Google don't exhaust the topic you're interested in, you've probably made a wrong query. Check your typos. And it may even get frightening when AI algorithms provide answers at a level of complexity that is far beyond our comprehension. For example, DeepMind AlphaZero exhibits vastly superhuman performance at chess, shogi (Japanese chess) and Go, and learned all that in a few hours (OK, on vast Google tensor processing units, but still), only by self-play, without using any human knowledge. Its performance stuns grandmasters and suggests a *qualitative* improvement in the game play compared to the strategies invented by professional chess and Go players over centuries (Silver et al., 2017, 2018).

Nevertheless, the algorithms developed so far are still *narrow AI*, confined to relatively limited action spaces. If we want, we can switch them off and do things our way. They don't yet have the capability to expand the space of actions which are searched in order to maximize the objective. AlphaZero is vastly superhuman at chess, shogi and Go, but it is unable to go from there and beat the stock market or master nanotechnology. It can't push its constraints, self-rewrite and go rule (let alone, navigate) the world. But we shouldn't exclude the possibility that one day an artificial general intelligence (AGI) might be able to do that.

The quest for AGI opens the threat that machine intelligence will one day obtain the capability to search a *very broad* space of actions while maximizing some objective which to some (even tiny) extent opposes our human local control. By the logic of intelligence explosion via a cascade of recursive self-improvements (Yudkowsky, 2013; Hanson and Yudkowsky, 2013; Bostrom, 2014), this AGI may quickly develop itself to be orders of magnitude faster and more accurate at decision making than people. We are, in the end, limited by the cognitive capacities of our brains that we cannot rewire, and victims of our puny data communication abilities based only on speech and writing. (Just compare that to a modern broadband Internet connection.) Then, one day we may realize that the objectives we had programmed into the AGI have led to unintended consequences which cannot be corrected anymore: the AGI would be already superintelligent and would always outsmart the humans, successfully protecting its original objective. In this way, we may fall victim to our own incompetence, myopia and dynamic inconsistency of actions: the AGI may get out of hand, giving the humankind a very narrow (or zero) time margin for reaction.

As an economist I expect that if one day AGI becomes technically possible, it will be built. Whether a technological corporation, government agency or an open-source software community, someone will eventually do it—and thereby become rich, famous and powerful, with lots and lots of local control. Economic incentives for developing AGI are crystal clear. After all, the biggest fortunes in the world are already made in the software business, and developing machine intelligence is a big growth driver there, as evidenced by the global position of Google, Facebook and Amazon.

So, will the rise of AGI constitute the next filter in the development of the human civilization? There are at least two reasons why it may indeed be. First, the *orthogonality thesis*: any level of intelligence can be coupled with more or less any final goal (Bostrom, 2012). Thus, as AI gradually approaches the human intelligence level we must be *extremely* careful in choosing the goals of our strongest AI algorithms. If an intelligence explosion (hard AI takeoff) occurs and we get the goals wrong, we'll never get a chance to correct them. In this regard, Nick Bostrom devised a fascinating but deeply disappointing doomsday scenario of *apocalypse by paperclips*. It goes as follows. Imagine that a group of AI researchers have come up with a new promising AI architecture but are not sure how well it will work. They thought that the best way of checking how strong it is, is by letting its first practice on a trivial goal that is easy to monitor. So they said, OK, for starters let's try the following: think of a most efficient way of making paperclips. They plugged the computer into the Internet, switched the program on and went to sleep. But they badly underestimated the power of their new algorithm. Overnight it managed to rewrite its code multiple times, dramatically improving its own performance, and learned how to harness the computing power of most devices connected to the Internet. In the morning no human on Earth was able to switch it off anymore. But the algorithm assessed that disabling the off switch is not enough. It realized that the most important security threat, with the potential to compromise its goal, is due to the *homo sapiens*. Logically, it decided that security threats must be eliminated. At the same time, it also began to—conscientiously, ingenuously and unhumanely quickly—turn the Earth into a huge laboratory aimed at perfecting the art of making paperclips.

The orthogonality thesis implies that we can't be comforted that the AI would act like a human once it reaches human-level intelligence. Most likely it won't. Moreover, the AI doesn't need to gain consciousness or ability to reflect on its own goals before it can become dangerous to humans. These are antropomorphizations characteristic for science-fiction movies. In reality, one should rather expect some variant of apocalypse by paperclips. The challenge of building a safe AGI is really huge and involves, in particular, the necessity of including human prosperity, well-being, and *everything that we value* directly in the AI's goal. Behavioral rules and optimization constraints are useless in this regard because a smart algorithm will always find its way around them, for example by adding new dimensions and finding surprising re-interpretations of ambiguous concepts (Hanson and Yudkowsky, 2013). Another major problem is that the goal has to remain invariant under recursive self-improvements of the algorithm. And it doesn't help that we don't quite know the goals of our lives yet.

The second reason why AGI can be a filter to the humankind lies in the sheer power that comes with intelligence. It is certainly healthy to be cautious about voluntarily setting ourselves in the position of an entity with inferior intelligence. After all, what was the fate of other hominins when confronted with the (slightly) more intelligent *homo sapiens*? Well, they're not with us anymore! The Neanderthals and Denisovans have become extinct even though their relative disadvantage in brainpower and social skills was probably small. Then it doesn't take much imagination to believe that a superintelligent AI will resist our attempts at switching it off or changing its goals, and it will possibly decide to eliminate us altogether, if not through deliberate genocide then for example by depriving us of the resources that we need for survival but it needs for building its paperclips.

At this point it is worthwhile to reflect on the historical dynamics of human local control. No matter how we measure it (and I will deal with this issue in much more detail in Part III of the book), it is *explosive*, super-exponential—that is, its growth rate is itself growing over time. This dynamic began right at the moment when the human local control maximization process emerged as a runaway sub-routine of the grand process of species evolution. It escaped evolution's control because of its pace, recursive self-improvement and greed for resources. The resulting cumulative technological and economic development eventually brought the humankind to a point where we could begin experimenting with AI algorithms which themselves—as an instrumental goal—tend to maximize their own local control. So far it remains innocent because the optimization power we could generate has remained way below the broad-based intelligence of the human brain. But the progress in AI research is fast and will only accelerate once the algorithms learn not just how to fine-tune their parameters but also how to redesign their own architecture. An ensuing intelligence explosion—if it happens—will reduce us to mute observers of how the AGI escapes from our control, for exactly the same reasons that humans fooled the evolution of species: (i) pace, (ii) recursive self-improvement and (iii) greed for resources. (Another variant of this prediction is the concept of a Malthusian population race between humans and artificially intelligent agents, laid out by Korinek and Stiglitz (2019).)

And there is one final thought you may deliberate with your friends and family: what's a filter, or *existential risk*, for the humankind, need not be a filter for the human civilization. It is conceivable that an AGI may carry human accomplishments into the future, only without our biological species being able to observe it. It may build on the products of our expansion, economic growth and technological progress, and convert them to serve its own (human-programmed) objectives, of which we may or may not be proud. What do you think?

References

Barabási, A.-L., & Albert, R. (1999). Emergence of scaling in random networks. *Science, 286*, 509–512.

Bostrom, N. (2012). The superintelligent will: motivation and instrumental rationality in advanced artificial agents. *Minds and Machines,22*, 71–85.

Bostrom, N. (2014). *Superintelligence: paths, dangers, strategies*. Oxford University Press.

de la Croix, D., & Dottori, D. (2008). Easter Island's collapse: A tale of a population race. *Journal of Economic Growth,13*, 27–55.

Diamond, J. (1997). *Guns, germs, and steel: The fates of human societies*. Norton & Company.

Dunbar, R. I. M. (1998). *Grooming, gossip, and the evolution of language*. Harvard University Press.

Hanson, R. (1998). *The great filter—are we almost past it?* (Online manuscript). George Mason University.

Hanson, R., & Yudkowsky, E. (2013). *The Hanson-Yudkowsky AI-Foom Debate*. Machine Intelligence Research Institute.
Harari, Y. N. (2014): *Sapiens: A brief history of humankind*. Vintage.
Kolbert, E. (2014): *The Sixth Extinction: an unnatural history*. Henry Holt and Company.
Korinek, A., & Stiglitz, J. (2019). Artificial Intelligence and Its Implications for Income Distribution and Unemployment. In A. Agrawal, J. S. Gans & A. Goldfarb (Eds.), *The economics of artificial intelligence: An agenda*. University of Chicago Press.
Silver, D., Hubert, T., Schrittwieser, J., et al. (2018). A general reinforcement learning algorithm that masters chess, shogi, and Go through self-play. *Science,362*, 1140–1144.
Silver, D., Schrittwieser, J., Simonyan, K., et al. (2017). Mastering the game of Go without human knowledge. *Nature,550*, 354–359.
Twenge, J. M., & Campbell, W. K. (2009) *The narcissism epidemic: Living in the age of entitlement*. Simon & Schuster, Inc.
Yudkowsky, E. (2013). *Intelligence explosion microeconomics* (Technical report 2013-1). Machine Intelligence Research Institute.

Part II
The Growth Mechanism

Chapter 4
Hardware and Software

A machine broke down in an assembly line of a large production facility. Further production was impossible. The equipment was costly, and so was each hour of assembly line downtime. A specialist was immediately summoned to the spot. He carefully examined the machine and identified the problem. Then he took a hammer and hit the machine hard, two times in the same place. Then he pressed the start button. The machine started working immediately. Everyone sighed with relief.
– Thank you. So, how much would it cost?—asked the production manager.
– $1000 – was the answer.
– A thousand dollars?! For two hits of a hammer?!—the manager was shocked—In such case, you must prepare a detailed invoice with all necessary repair actions.
The repairman just shrugged his shoulders. On the next morning the manager found an invoice on his desk, saying:
Hitting the machine with the hammer—$2.
Knowing where to hit—$998.

random Internet source, own translation

4.1 A New Model of Growth

The humankind pursues local control maximization by performing premeditated actions, the sum of which sets the course of development of the human civilization. The macroeconomic way of looking at it assumes an aggregate *production function*, describing how inputs are transformed into aggregate output. Throughout most macroeconomics literature, the default choice of production inputs has been capital and labor, and the default definition of output has been value-added or GDP. In models of an agricultural economy, in turn, the inputs are land and labor, and the output is the volume of agricultural production. Furthermore, in some studies the labor

J. Growiec, *Accelerating Economic Growth*, Frontiers in Economic History,
https://doi.org/10.1007/978-3-031-07195-9_4

input is considered in effective units, reflecting differences in education and skills. All in all, growth mechanisms discussed in the existing economics literature tend to be restricted to one particular era only—typically the industrial era—or a transition period between two eras—typically the industrial revolution. As such, they are largely insufficient to handle all technological eras simultaneously.

All eras can be handled simultaneously within my *hardware–software framework* (Growiec, 2019, 2022b), though. The fundamentals of that framework come from physics. My key assumption is that in any conceivable technological process output is generated through purposefully initiated physical action. In other words, producing valuable output requires both some physical *action* and some *code*, set of instructions describing the action. I therefore postulate a general production function featuring two inputs: physical *hardware*, performing the action, and disembodied *software*, providing information on what should be done and how.

Hardware and software are mutually *complementary* and each of them is *essential* for production. Complementarity (or more precisely: gross complementarity) means that an increase in the use of one factor increases the output elasticity of the other factor (Klump et al., 2007, 2012). Intuitively, it means that factors must be used together: the more we increase the imbalance between them, the more it pays off to bring the balance back to normal. The more machines you install at a factory, the more beneficial it is to make them operative, for example by employing staff able to feed them with appropriate instructions.

Essentiality, in turn, means that you simply can't produce any output without one of the factors. An idle machine that doesn't receive any instructions, or free-floating code that isn't run on any device, produces nothing.

In the extreme case of perfect complementarity, factors must be used in fixed proportions and any surplus ("leftover") factor is wasted. Cooking recipes, for example, begin with a list of ingredients to be used in specified proportions. If you fiddle with them, you will cook something else than expected and probably you won't be satisfied with the result. However, you may want to choose your preferred technology from a broader technology menu, and then the complementarity at the aggregate level will be less than perfect even if each technology is a recipe (Jones, 2005; Growiec, 2013, 2018). If your goal is to cook a good dinner, for example, you are not tied to a single recipe but may choose from a range of recipes, each of which uses ingredients in different proportions, and then you may fiddle with factor proportions indirectly by making the choice of recipe. There is just one way to cook a perfect *Spaghetti Bolognese*, but many ways to make a great dinner, including ones that don't use any tomatoes.

The same logic can be applied at the macroeconomic level to hardware and software. Aggregate output (say, for example, GDP) may be produced in many ways, depending on the choice of production technology. Imperfect factor complementarity means that producing output requires to use both hardware and software in the process but there is a possibility of generating the same outcome with more physical action but less efficient code, or vice versa. Essentiality of hardware and software, however, restricts the choice of options so that nothing can be produced without any physical action or without any code.

An important simplifying assumption of the above-specified hardware–software model is that it abstracts from raw materials and energy which are being used up in the production process (Georgescu-Roegen, 1971). The model works as if raw materials (from natural resources) and energy (derived from burning nutrients in our body cells, photosynthesis in plants, or generated in engines and power plants) are given for free and in practically unlimited supply. Although we can rely on the Sun to bombard our planet with free photons for the next 5 billion years or so, this assumption is not exactly verified, though, and one-day scarcity of natural resources and energy may become a constraining factor in further development. Indeed, that's one of the key warning signals from Club of Rome reports (Meadows et al., 1972, 2004) and the associated voluminous literature on exhaustible resources. Still, in my view, human ingenuity stemming from the local control maximization has so far managed to keep this issue at bay and will continue to do so in the foreseeable future. Throughout the history, it was not the availability of resources that limited the extent of our control but the low efficiency with which our hardware and software transformed them into the desired output.

4.2 What's in Hardware, What's in Software

But what do I mean exactly by hardware and software? Let's discuss some important specifics.

Hardware includes all agents of physical action, and therefore encompasses human physical labor, the labor of domesticated animals, agricultural land, as well as all sorts of productive machines (or as economists would say, physical capital). The concept requires that all physical labor included in hardware must exclude any know-how or skill of the worker, and hardware services of programmable machines must exclude their code. (In this regard, note that although contemporary programmable machines are mostly computers and robots featuring chips of integrated circuits, in principle they don't have to be silicon-based. In fact, the first pieces of "programmable hardware" in history were mechanical devices such as the Jacquard loom using punchcards, first invented in 1804. Who knows, perhaps in the future we will learn to program, e.g., nanobots or artificial organic beings?).

Software, in contrast, encompasses all the instructions used for performing the physical actions. Throughout the human history, they have been the outcome of human cognitive work, and therefore stemmed from the human know-how, ability to use state-of-the-art technologies and willingness to cooperate. Nowadays, however, it also increasingly includes pre-programmed software, understood as a task-specific list of instructions to be performed by the associated programmable hardware, as well as artificial intelligence (AI) algorithms, i.e., programmed software which is able to learn from data as well as potentially self-improve and self-replicate.

Within hardware, the agents of physical action are substitutable. This reflects the idea that whatever performs a given set of actions, if the actions are precisely defined then the outcome should be the same. The same logic applies to software: regardless

Table 4.1 Factors of production

Hardware	Human physical labor
	Physical labor of domesticated animals
	Agricultural land
	Non-programmable physical capital
	Programmable physical capital
Software	Human cognitive work
	Pre-programmed software
	AI algorithms

of whether a set of instructions comes from a human brain or a digital information processing unit, if the actual information content of instructions is the same, then the outcome should be the same, too. A qualification to this general regularity is that at a given level of technological advancement some types of hardware or software may be indispensable for certain actions. For example, neither agricultural nor industrial technology developed so far allow to substitute for land in producing agricultural products. Similarly, the digital technologies of 2022 don't yet allow to replace human cognitive work in many sophisticated tasks involving creativity and judgment.

Each of the specific factors of production (see Table 4.1) has its unique properties. In particular, they differ crucially in the degree of *rivalry* (Romer, 1990). Human work, both physical and cognitive, as well as land and physical capital, are rivalrous, which means that they can be used for production only once at a time. If you use them in process A, you cannot simultaneously use them in process B. Knowledge, technology and computer code are—in contrast—non-rivalrous and may be used in many processes at the same time. To build a house, you need bricks and mortar as well as the physical labor of a building crew, and once you contract these inputs they cannot be simultaneously used for building your neighbor's house. The idea of how houses are built, or even a precise project of the house, is however not "used up" in the process: your neighbors may, if they like, build their house following exactly the same project.

The non-rivalrous component of software is disembodied *information* (Romer, 1990; Hidalgo, 2015). Information in the form of knowledge or skills can be passed from one human brain to another without compromising the quality of the original—parents and teachers do this routinely—and thus has the potential to scale up with the human population. By the same token, the code of pre-programmed software and AI algorithms can be virtually costlessly copied from device to device and thus has the potential to scale up with all programmable hardware. (Note by the way that the concept of rivalry refers only to hard physical constraints rather than legal or economic restrictions, such as intellectual property rights and software licensing, that are a matter of agreement.)

We can be even more specific with regard to the properties of hardware and software. Namely, within hardware, human physical labor is rivalrous and given in fixed supply per capita and per unit of time. This reflects the dire reality that we can't do bilocation, are very bad at multitasking, and that our day lasts always 24 h, of which roughly one-third is spent asleep. Labor of domesticated animals is also rivalrous and animals are subject to similar life constraints. A key difference is that, looking from the human's perspective, animals can be accumulated. You have just one body, but you may own as many horses as you can afford; "per capita" doesn't include horses. Agricultural land is also rivalrous and also can be accumulated per capita. Its total supply is however firmly fixed at the global level by the Earth's landmass. Physical capital, encompassing all structures and equipment—all sorts of devices that can be used for generating output—is also rivalrous, but in contrast to land, can be accumulated virtually unboundedly. In further contrast to animals, capital accumulation is measured in value terms and reflects not just quantity, but also quality changes in the productive inputs. Furthermore, physical capital may be non-programmable (like a factory building or a shovel) or programmable (like a computer or a robot).

Within software, human cognitive work refers to the sum of mental processes happening in human brains that provide instructions for productive action. It is a product of three factors: available technological knowledge, individuals' skill levels and the number of workers. Technological knowledge, or the size of the "repository of codes" available to people, is non-rivalrous and accumulable. The per-capita skill level is also non-rivalrous, but costly to transfer and bounded above—theoretically by the optimal code for performing a given task, but in practice by a much lower number due to (i) human mortality which implies that each generation must learn their skills anew, and (ii) limitations of human brainpower, coupled with our inability to rewire our brains in order to perform cognitive tasks more efficiently. The number of cognitive workers is of course rivalrous and in fixed per capita supply: each of us has only one brain. (Observe that all productive tasks that people perform require both the bodily hardware and neural software. Mundane physical tasks like digging a ditch are hardware-intensive but still require some cognitive work to coordinate muscles on the goal. At the other end of the spectrum, even extremely software-intensive tasks like philosophical reflection rely on computations carried out in the brain, and so they need the bodily hardware, too. In fact, an adult human expenses about 20% of daily energy on the functioning of her brain, even though it constitutes on average only 2% of her mass, Raichle and Gusnard (2002).)

Pre-programmed software and AI also consist of three multiplicative factors: technological knowledge, "AI skill level" which captures the degree and efficiency with which the software is able to perform the tasks, and the stock of programmable hardware on which the software is run. I assume without loss of generality that overall technological knowledge (volume of the "repository of codes") available to humans, pre-programmed software and AI are the same. (Any discrepancy can be easily reflected in the relative skill levels of human cognition and AI.) The AI skill level is non-rivalrous and bounded above by the optimal code for performing a given task (e.g. perfect accuracy), though there may be in fact a much lower upper bound (Hanson and Yudkowsky, 2013). The stock of programmable hardware on which the software is run is rivalrous and accumulable per capita.

The hardware–software model envisages technological progress as expansions of the "repository of codes", i.e., as the development of new, better instructions allowing to produce higher output with a given amount of hardware. These instructions can be applied to any task at hand both by humans, deterministic pre-programmed code, and AI. This is intuitive: new technologies may come in the form of abstract ideas, scientific theories, systematically catalogued facts, codes specifying certain actions or blueprints of physical items. All this is *information* and not actual *objects* or *actions*, and it is precisely this informational character that makes ideas non-rivalrous (Romer, 1990). In this way my model resolves the (at least 60 years old, Solow (1956)) conundrum of whether technical change increases the productivity of capital or labor. In fact, it targets neither of them directly: what it affects immediately is software. (In economic jargon we would say that technical change is *software-augmenting*.) This means that it augments cognitive but not physical labor and programmable capital to the extent it accommodates pre-programmed software and AI.

4.3 What Outputs Do We Produce?

It is also worth reflecting on the *outputs* of our productive actions. In this regard, macroeconomics has long cherished the tradition of lumping all sorts of outcomes into a single final output variable Y such as value added or gross domestic product (GDP). The broadly accepted System of National Accounts offers a detailed methodology of computing GDP while maintaining comparability across time and between countries. For international comparability, economists adjust GDP for differences in purchasing power across countries; for intertemporal comparability, we build real series that are corrected for changes in prices. Macroeconomics also emphasizes that final output consists of consumption goods and investment goods so that part of our output serves to satisfy our human needs directly, whereas the remainder is invested and used to improve our ability to generate consumption goods in the future.

Economic growth theory, however, ever since the Nobel prize worthy contributions of Paul M. Romer (1986, 1990) has also considered a second, distinct output variable: ideas, or new technological knowledge, produced in the process of research and development (R&D). Models following this observation, which we call R&D-based endogenous growth models, have thus always featured (at least) two aggregate production functions, one for value added and one for ideas (Gomułka, 1990; Aghion and Howitt, 1992; Jones, 1995, 1999, 2005; Ha and Howitt, 2007; Madsen, 2008; Kruse-Andersen, 2017; Bloom et al., 2020).

Another type of output which has been considered in some long-run growth models, separately from value added and ideas, is human capital, also referred to as the average skill level in the population or educational attainment. It depicts the functioning of the education sector where the inputs such as teachers' time and skills as well as students' time, or overall education expenditures, are transformed into increases in human skill levels, allowing the graduates of the education system to perform cognitive work more effectively (Uzawa, 1965; Mincer, 1974; Lucas, 1988;

Table 4.2 Typology of final outputs

Consumption goods	Agricultural produce
	Manufactured consumption goods
	Consumer services
	Cultural goods
Hardware accumulation	Seeds for planting and animal feed
	Manufactured capital goods
Software accumulation	New technological knowledge
	Education
	New computer software and AI algorithms

Heckman et al., 2003). R&D and education, as well as computer programming, are different from the creation of general value added and merit separate production functions because their output—new technological knowledge, human skills and digital code—is non-rivalrous.

Table 4.2 presents a typology of final outputs, split into three categories: *consumption goods*—outputs that serve the short-run goal of satisfying human needs; *hardware accumulation*—tangible outputs that are invested in order to have more hardware available in the future; and *software accumulation*—intangible outputs that add to the available software.

Of particular interest are the outputs specific to the digital era. Some of them are cultural consumption goods: computer and video games, Internet videos and other forms of digital entertainment satisfy our consumption needs, just as paintings, sculptures, live music and theatrical performances did over the centuries. Other digital outputs, in turn, add to software accumulation: new computer software and AI algorithms add to the "AI skill level" just like education adds to our human skill level.

The hardware–software model also carries an important message on the role of *data*. Namely, raw data is to software what raw materials and energy are to hardware: an essential input that is subject to transformation. Purposefully initiated physical actions use natural resources and expend energy, leading to creation of valuable output. Similarly, in the realm of computation purposefully initiated cognitive processes may use raw data (and also expend energy) in order to generate useful information.

4.4 Hardware and Software in Pre-industrial Eras

Back in the hunter–gatherer era, everything was simpler. Humans devoted most of their lives to the search for food. Their hardware was their brawn and their software was their brains. The output they generated was total food production from habitats under human control. The natural habitat capacity, securing the flows of food and other natural resources (e.g. furs for clothing, wood for shelter, healing herbs, etc.) from a given piece of land, was essentially fixed. All production was immediately consumed. Population density was determined by the carrying capacity of the ecosystem and the Malthusian mechanism made sure that the "quality of life" (in those times, equivalent essentially to food consumption) was fixed at the subsistence level. Any increase in output meant that additional calories were available that could feed more people, and the population adjusted accordingly.

There were only two mechanisms of development in the hunter–gatherer era. First, territorial expansion: as humans conquered additional habitats, they were able to increase food production and thus the overall population size. Second, accidental discoveries and conscious experimentation that led to better knowledge about the characteristics of plant and animal species, better stone tools, hunting methods, new uses of fire etc. The former followed from the instrumental drive of resource acquisition whereas the latter from the drive for creativity and technological perfection. The latter also resulted in a (very slow and gradual) buildup of "technological knowledge", the non-rivalrous component of software (Diamond, 1997; Pacho, 2013).

However, this new technological knowledge was not matched by proportional increases in output because production was constrained by the scarcity of physical labor and limited by the carrying capacity of the natural ecosystem. Luckily, there came a period in history when human knowledge about various plant species as well as the available skill set became sufficient to allow first groups of people—located in the Fertile Crescent covering parts of today's Syria, Iraq and Turkey—to plant and harvest wheat, barley and a few other crops. Success in domesticating first food crops (as well as some large herbivores) led to a gradual transition to sedentary agriculture. At this point, the humankind learned to "hack" natural growth processes and coerce them into providing energy for our own species. If the cognitive revolution—kickstarting cumulative knowledge accumulation—was our first blow against the evolution of species, the agricultural revolution was a knock-down: instead of adapting our bodies to the requirements of the environment, we began adapting the environment to requirements of our bodies. What followed was the *MacCready explosion*: at the dawn of agriculture, the worldwide human population plus livestock and pets was approximately equal to 0.1% of the terrestrial vertebrate biomass; today it is about 98% (Dennett, 2017).

The human ability to transform natural habitats into farmland added a second component to our hardware: agricultural land. A third component—animal labor—was simultaneously added, thanks to the domestication of animals. The inclusion of both components in hardware not only resolved the former scarcity of this input but also markedly accelerated the pace of development. This is because, compared to the previous hunter–gatherer era, the agricultural era strongly increased the production of calories per land area—which, in a Malthusian world, translated into a respective increase in population size. It was not a one-shot improvement; however, the overall area of cultivated land gradually increased over time, and so did total livestock, and in consequence also the total human population.

The agricultural economy also opened a new dimension of economic development, namely the increases in efficiency of agricultural production (yield per land area). In contrast to hunting and gathering whose output was limited by the capacity of the natural ecosystem, farming showed massive potential for yield improvements. Moreover, increased population density and the availability of food surpluses—even if temporary—allowed for centralization of power and creation of state. Consequences of these developments ranged from monopolization and thus overall reduction of violence, easier diffusion of ideas after the invention of writing, to the emergence of first empires, non-food-producing specialists as well as narrow, wealthy land-owner elites which, if they wanted, could afford to be well-educated and nurture their inquisitiveness and creativity for example by writing philosophical texts (Kremer, 1993; Diamond, 1997; Hibbs and Olsson, 2004; Galor, 2011; Borcan et al., 2018).

The transition from hunting and gathering to agriculture took millennia to complete. Due to differences in climate across world regions and specific requirements of food crops, diffusion of agricultural technologies sometimes stopped, and in fact agricultural revolutions appeared *independently* in at least seven regions of the world (Diamond, 1997). The average pace of development of agricultural economies was—by today's standards—very slow, and periods of development alternated with episodes of stagnation or decline. It is not inconceivable that the agricultural era could have prevailed much longer if not for the events that took place in Europe between the fourteenth and sixteenth century CE.

After the pandemic of Black Death killed between one-third and one-half of the Western European population in 1347–1351, the region gradually re-emerged with a modified social structure. Depopulation further eroded the peasants' already weakened obligations and allowed them to implement labor-saving innovations, such as a shift from grain farming to animal husbandry. Increases in the marginal product of labor also allowed some of the former peasants to migrate to cities, earn higher wages and improve their life standards. This period witnessed, moreover, a gradual expansion of international trade and a sequence of great geographical discoveries and first European colonial conquests. When land became relatively abundant, production became constrained by the level of technological sophistication of human work. *Software* has become scarce, and as we know, scarcity is a mighty force that coordinates human efforts on overcoming it.

Once diffusion and storage of knowledge and information were greatly facilitated by Johannes Gutenberg's fifteenth-century invention of the printing press, and wealthy elites were willing to invest in the development of universities, conditions were set for a new breakthrough in software—the scientific revolution. During the next two centuries Europe witnessed the rise of modern science, including empiricism, experimentation, mathematization and hypothesis testing. An environment was created that systematically facilitated the nurturing, spreading and further development of brilliant new ideas. Never before has philosophy and science been so successful in explaining the world and opening new development opportunities. As a result, technological progress was no longer an accidental effect of more or less random tinkering by a few privileged ones but has become institutionally supported and performed on a full-time basis by some of the more gifted Europeans (though, regrettably, still recruited predominantly among noble men).

The take-off of science and technological progress during the European "enlightenment era" (sixteenth to eighteenth century CE) implied that soon it was not software which was scarce in production, but—once again—hardware. Improved efficiency of human cognitive work led to Europe's global dominance in terms of economic power and technology, including warfare. This enabled Europeans to go wild in their drive for resource acquisition. In the following centuries, European empires colonized most of the world, ruthlessly exploiting its populations and natural resources. In the Americas and Australia, after Europeans' "guns, germs and steel" wiped out most of the indigenous peoples, colonial powers set up agriculture with Eurasian food crops and domesticated animals. Soon these lands were inhabited by ever-increasing populations of European descent (Diamond, 1997; Galor, 2011).

Despite all that territorial expansion hardware remained scarce relative to software during the enlightenment era because as opposed to land and labor, technological knowledge is non-rivalrous and almost costlessly copied. Hence, the remarkable success in "extensive" resource acquisition after the scientific revolution was accompanied by rather limited growth at the "intensive" margin. European empires grew fast in territory, population and total output, but not that much in output per worker. Further progress in this regard required a new breakthrough in hardware.

4.5 Engines of Modern Growth

Inventions of the industrial revolution paved the way for multifold increases in the human's ability to harness energy and transform it into productive work. The steam engine, and then electricity and the internal combustion engine were key milestones in the development of humankind because they dramatically increased our capacity for fulfilling our needs and increasing our local control.

Technologies of the industrial era were first introduced around 1800 CE in England and soon adopted and developed across Europe. From that point onwards the humankind began to accumulate *physical capital*—various mechanical devices able to perform physical actions according to human instructions. Our stock of hardware skyrocketed because these new fossil fuel-propelled machines were able to generate output much more efficiently than agricultural land and human physical labor. In the early years of the industrial revolution, physical capital was relatively scarce and easy to accumulate. Capitalists began to build factories, manufactures, assembly lines and set up their enterprises. They also quickly figured out the division of labor and gauged it into increasing returns to scale (Galor, 2005; Gordon, 2016). All this vastly increased the overall efficiency of production.

The consequence of the fundamental complementarity between hardware and software, though, so badly missed by Karl Marx and his disciples, was that the industrial revolution in hardware actually made software increasingly scarce. Instead of capital owners endlessly turning the screws on the working class, after the first turbulent decades of the nineteenth century labor laws were introduced, and the hardest and most repetitive physical tasks were mechanized. Then the economy set on to increase its demand for the services of cognitive, skilled labor. (Economists identified this broad regularity only a century later. We call it *capital–skill complementarity*, cf. Krusell et al. (2000).) Gradually, employees' cognitive work was becoming more and more valued relative not only to their physical efforts but also to the services of capital.

Soon also directed policy actions were taken to counteract the mounting scarcity of human cognitive work. The big yet untapped pool of "software" was the cognitive work that could be performed by the previously uneducated lower classes. In response, public education was introduced. Soon elementary education increased the skill levels of the general population, whereas vocational education acquainted them with some of the most useful contemporary technologies. All this mounted up to secular increases in *human capital* per worker. The equilibrium response of the economy was an upward trend in average wages. During the industrial era, the capital share of output and income inequality first went up (in the Marxian period) and then down (Galor, 2005, 2011).

But the industrial revolution was not just about mechanization, education and social reform. The scientific developments that underlay the dawn of the industrial era included, besides the new ways of harnessing energy, also major breakthroughs in medicine and agriculture. Largely reduced infant mortality and rising food supply, thanks to overseas farmlands and greatly improved agricultural productivity, marked the onset of a demographic explosion. The world population began to rise super-exponentially, first doubling between 1750 and 1900 CE (from ca. 0.8 to 1.6 billion) and then reaching 6 billion by 2000 CE. Several decades later, growing economic returns to skilled cognitive work, coupled with the growing availability of birth control, activated the *children quantity–quality tradeoff*: parents gradually limited the number of their offspring while increasing their investment in each of them. Reduced fertility eventually alleviated the Malthusian mechanism of convergence to subsistence levels of consumption, allowing the humankind to achieve permanent, broad-based and cumulative increases in standards of living.

Mature industrial economies of the late nineteenth and early twentieth century (up to the 1970s) were characterized by Kaldor's (1961) "stylized facts". In that period, factor shares of output were roughly constant and so was the return to capital. Wages, in contrast, exhibited a steady exponential upward trend. Capital, wages, consumption and output grew at roughly the same pace. The fundamental driving force of long-run economic growth was technological progress (sustained growth in the non-rival "repository of codes"), further accompanied by marked increases in physical and human capital per capita over a prolonged transition period (Barro and Sala-i-Martin 2003; Jones 2005). It was a period of unusual, fragile balance between growth in hardware (physical capital) and software (population size, average educational attainment and technology). This regularity, coupled with the lack of reliable numerical evidence for earlier centuries, made many twentieth century economists believe that Kaldorian *balanced growth* is a law of nature that will stay with us forever, and that the times of factor scarcity are finally over.

However, with the wisdom of hindsight we can now confidently say that not only was this balance absent before the industrial revolution (Galor, 2011) but also that Kaldor's "stylized facts" ceased to hold from the 1980s onwards (Jones and Romer, 2010). In the 1980s, the human civilization entered into the digital era, characterized by a few unmistakable symptoms. The labor's share of global output began to fall (Karabarbounis and Neiman, 2014), firms' average profit margins and CEO pay went up (Gabaix and Landier, 2008; Barkai, 2020), and within-country income inequality increased, particularly at the top of the income distribution (Piketty and Saez, 2003; Piketty and Zucman, 2014; Jones and Kim, 2018). Growth in our capacity to process, store and communicate information has decoupled from growth in GDP and capital investment, and data processing has decoupled from the capacities of the human brain. The digital computer has become the new steam engine and data have become the new oil.

What irreversibly undermined the shaky balance between hardware and software at the dawn of the digital era was the creation of *programmable hardware*. It worked in a perverse way: computers, despite being a development to hardware, by being able to store and run their code actually opened an avenue for disproportionately accelerated accumulation of software. For the first time in history, human instructions could be taken out of the loop. New forms of software included digitally stored useful data, pre-programmed algorithms and—soon after—AI algorithms that are able to learn from data on their own. This swung the long-run developmental pendulum once again toward software.

Programmable hardware of the digital era includes all sorts of digital devices (computers, cell phones, smartphones, industrial robots, smart refrigerators etc.), whereas pre-programmed software involves the algorithms and programs which are run on these devices. Nevertheless, the broad "software" amalgamate still includes, to a huge albeit declining extent, human cognitive work. As of 2022, it's still human programmers who ultimately write the instructions to be performed by computer hardware, and plenty of cognitive tasks are still performed without any programmable devices whatsoever.

Recall, however, that physical hardware is rivalrous while information is not. Computer software and AI can be almost costlessly copied, and thus if not for legal or economic constraints, it has the potential to quickly scale up to the available programmable hardware. Thus far, in the absence of globally coordinated data policy, this scalability has led software companies such as Microsoft, Google and Facebook to become global superpowers, and their founders—Bill Gates, Larry Page, Sergey Brin and Mark Zuckerberg—to be the richest people on the planet.

4.6 What's Coming Next?

So how will the digital era economy develop in the future? Economic logic as well as historical analogies suggest that we are still in the early days of the digital era. There is more computing power than we really need given our software, and there is more data on our hard drives and server clouds than we can process. Today's computer software generally doesn't use the available computing power efficiently because it remains specialized and rigid in terms of data requirements, and for the most part it still relies on the slow interaction with human users. Human-centered software features such as pretty graphics or intuitive frontend solutions take away a lot of processor power and burden memory with information which is not important for the core digital process ("what Intel giveth, Microsoft taketh away.") In effect, despite the tremendous growth in computational capacity and communication bandwidth, the level of automation of the digital era production processes remains low. Most of the time when computing power stays idle is when the user has to set the instructions manually. Despite all the progress, software is still a scarce factor and the bottleneck of economic growth.

Because computing power is accumulated at the pace of Moore's Law (doubling every two years or so, Hilbert and López (2011)) and software isn't, the price of software relative to hardware is quickly going up. So far this has been mirrored in a growing skill premium: wages of skilled individuals able to perform cognitive work complementary to the expanding and improving programmable hardware have been growing disproportionately fast. For example, the IT sector offers some of the highest wages these days, and paychecks offered by the digital giants of GAFAM (Google, Amazon, Facebook, Apple and Microsoft) allow them to compete for the world's top talent. However, Moore's law in hardware also creates potentially huge economic rewards to developing digital software such as AI, that would replace humans in certain tasks. The promise of automation is to circumvent the slow human thinking and even slower computer–human communication and replace them with within-computer computation and computer–computer data flows that are orders of magnitude faster (and also more accurate). Accordingly, largest increases in productive efficiency are observed when tasks are fully automated.

Although there has been a place for both human minds and computer algorithms in the "software" amalgamate so far, it doesn't have to always be the case. Existing algorithms lack the cross-domain versatility of the human mind (Yudkowsky, 2013) and the ability of ideation and creativity (Brynjolfsson and McAfee, 2014), but in the confines of their application they are vastly faster and more accurate. They are also getting better each day at pattern recognition based on big data, classification, categorization of various sorts of content and similar predictive tasks. Automation is already gradually eliminating routine jobs, both manual and cognitive (Acemoglu and Autor, 2011; Autor and Dorn, 2013; Frey and Osborne, 2017), and the remaining jobs are safe only until the development of sufficiently sophisticated and versatile AI technologies. Muehlhauser and Salamon (2012) present the key advantages of AI relative to humans: increased computational resources, vastly superior communication speed, increased serial depth (of thought), duplicability, editability, goal coordination and improved rationality. These advantages are important for gauging why AI is a potentially valuable substitute for people in performing the majority, if not all tasks of the digital era economy. If that's going to be the case, people may try to protect themselves against full automation of human tasks by creating new tasks and jobs (Acemoglu and Restrepo, 2018). Many of us are already entering, e.g., into creative niches of YouTube vlogging, Instagram modeling, videogame graphics art and so on. But will we be always able to do so?

The answer depends crucially on what we think about the cognitive differences between humans and machines. In this regard, despite foreseeable disagreement, I firmly hold that there is no qualitative difference between animal, human and computer thought. The entire difference is quantitative and stems from different levels of depth, complexity and versatility of data processing. This assumption implies that hallmarks of human proficiency, such as ideation, creativity and intuition, are just sophisticated forms of pattern recognition in which AI algorithms may one day "gradually, then suddenly" reach superhuman levels of performance (Brynjolfsson and McAfee, 2014). I am therefore embracing the possibility that even the most software-intensive tasks, like R&D, may be subject to automation in the digital era. In the future, AI may revolutionize R&D by not just answering research questions, but also asking new ones. (And AI is already used in, e.g., genome sequencing, processing of astronomical data, prediction of protein structures (Jumper et al., 2021), not to mention web browser engines, which are of enormous help to modern researchers.)

There are a few reasons to believe in that possibility. First, the capabilities of AI algorithms are apparently improving at a pace that is of the same order of magnitude as hardware accumulation (which is subject to Moore's law), Grace (2013), Hernandez and Brown (2020). Second, in the recent decade, we have witnessed a surge in breakthroughs of AI capability, ranging from autonomous vehicles and simultaneous language interpretation to self-taught superhuman performance at chess and Go, following from the same broad methodology of *deep neural networks*, deep learning (Tegmark, 2017). Third, this assumption just resonates very well theoretically with the principle of local control maximization as an emergent phenomenon arising spontaneously in data-processing systems.

In sum, my prediction for the coming years is that there will be a lot of automation—probably even a super-exponential trend in automation with no obvious upper bound below 100%. And there is no way around it because saving automatable jobs from automation is not only economically infeasible over the long run, but also irrational from the perspective of aggregate human control.

I also expect that in the coming years the falling share of human cognitive work in overall software will continue to contribute to increasing income inequality. A key driver of that trend is that in stark contrast to human cognitive work which is naturally dispersed—each of us has just one brain—computer software is almost costlessly copied, highly scalable, and sold to a single global market. In consequence, market concentration in the software business is extremely high (just think of GAFAM).

But I also expect that this trend won't go on forever—or otherwise I'd be repeating Karl Marx's error over again. Recall that over the course of the industrial era, income inequality first rose, fueled by the replacement of dispersed human physical work with concentrated physical capital, and then fell, fueled by the rising demand for complementary software, which was met with an increasing supply of dispersed human cognitive work. I don't see a reason why such a distinctive inverted U-shape pattern in income inequality (sometimes called the *Kuznets curve*) couldn't repeat also in the digital era. Currently—similarly to the times of Marx—within—country income inequality is on the rise, fueled by the replacement of dispersed human cognitive work with concentrated digital software. But, by analogy, this trend should reverse when software ceases to be the growth bottleneck: with sufficient software, the developmental pendulum will swing back once again and there will be again growing demand for hardware, met with growing supply. But at this point, digital hardware will be relatively more dispersed than software because of its rivalrous character, and therefore income inequality will go down. (This, of course, doesn't mean we shouldn't be implementing redistributive policies aimed at reducing income inequality. The level at which income inquality will "naturally" plateau may, and most likely will be socially unacceptable.)

Finally, the digital era may also end sooner than you expect. First, an existential risk may materialize and suddenly put the human civilization to its end. (In the wake of artificial general intelligence and advanced bioengineering the probability of this scenario is not entirely negligible.) Second, the next technological revolution may come along and once again revolutionize our technology, economy and society. Looking at the historical evidence of the developmental pendulum, I expect the next revolution ahead of us to augment hardware rather than software. Whether nanotechnology, quantum computing, synthesis of biological life, fusion power, qualitatively new solar power cells, or something else we can't yet imagine, it will be more likely to affect our ability to harness energy and perform physical actions than to accumulate and process information. (In the twentieth century there was one major breakthrough in energy generation: nuclear power. It didn't quite pan out as a new technological revolution in hardware, however, because its costs turned out to be in the same ballpark as the costs of other energy sources such as fossil fuels. Nuclear power plants are more efficient and environmentally less burdensome than power plants that burn fossil fuels, but at the same time also extremely costly to set up, plus they work on rare, difficult to handle and expensive fuel—uranium.)

In sum, whether there will be a new hardware revolution or not, technological and economic developments coming up in the next few decades will surely bring tremendous cumulative increases in human control. A question remains, though, how these gains—accruing to the collective brain of the humankind—will be distributed. And there is also the pressing question on limits to automation: will our inventions eventually make us redundant?

References

Acemoglu, D., & Autor, D. (2011). Skills, tasks and technologies: Implications for employment and earnings. In O. Ashenfelter & D. Card (Eds.), *Handbook of labor economics* (Vol. 4, Chap. 12, pp. 1043–1171).

Acemoglu, D., & Restrepo, P. (2018). The race between man and machine: Implications of technology for growth, factor shares and employment. *American Economic Review, 108*, 1488–1542.

Aghion, P., & Howitt, P. (1992). A model of growth through creative destruction. *Econometrica,60*, 323–351.

Autor, D. H., & Dorn, D. (2013). The growth of low-skill service jobs and the polarization of the US labor market. *American Economic Review,103*(5), 1553–97.

Barkai, S. (2020). Declining labor and capital shares. *Journal of Finance,75*, 2421–2463.

Barro, R. J., & Sala-i-Martin, X. X. (2003). *Economic growth*. MIT Press.

Bloom, N., Jones, C. I., Van Reenen, J., & Webb, M. (2020). Are ideas getting harder to find? *American Economic Review,110*(4), 1104–1144.

Borcan, O., Olsson, O., & Putterman, L. (2018). State history and economic development: Evidence from six millennia. *Journal of Economic Growth,23*, 1–40.

Brynjolfsson, E., & McAfee, A. (2014). *The second machine age: Work, progress, and prosperity in a time of brilliant technologies*. Norton & Co.

Dennett, D. (2017). *From bacteria to Bach and back: The evolution of minds*. Norton & Company.

Diamond, J. (1997). *Guns, germs, and steel: The fates of human societies*. Norton & Company.

Frey, C. B., & Osborne, M. (2017). The future of employment: How susceptible are jobs to computerisation? *Technological Forecasting and Social Change,114*, 254–280.

Gabaix, X., & Landier, A. (2008). Why has CEO pay increased so much? *Quarterly Journal of Economics,123*, 49–100.

Galor, O. (2005). From stagnation to growth: unified growth theory. In P. Aghion & S. N. Durlauf (Eds.), *Handbook of economic growth* (pp. 171–293).

Galor, O. (2011). *Unified growth theory*. Princeton University Press.

Georgescu-Roegen, N. (1971). *The entropy law and the economic process*. Cambridge, MA: Harvard University Press.

Gomułka, S. (1990). *The theory of technological change and economic growth*. Routledge.

Gordon, R. J. (2016). *The rise and fall of american growth: The U.S. standard of living since the civil war*. Princeton University Press.

Grace, K. (2013). *Algorithmic progress in six domains* (Technical report 2013–3). Machine Intelligence Research Institute.

Growiec, J. (2013). A microfoundation for normalized CES production functions with factor-augmenting technical change. *Journal of Economic Dynamics and Control,37*, 2336–2350.

Growiec, J. (2018). Factor-specific technology choice. *Journal of Mathematical Economics,77*, 1–14.

Growiec, J. (2019). *The hardware–software model: A new conceptual framework of production, R&D, and growth with AI* (SGH KAE Working paper No. 42). SGH Warsaw School of Economics.

Growiec, J. (2022b). What will drive global economic growth in the digital age? *Studies in Nonlinear Dynamics and Econometrics*, forthcoming.

Ha, J., & Howitt, P. (2007). Accounting for trends in productivity and R&D: A Schumpeterian critique of semi-endogenous growth theory. *Journal of Money, Credit and Banking,39*(4), 733–774.

Hanson, R., & Yudkowsky, E. (2013). *The Hanson-Yudkowsky AI-Foom Debate*. Machine Intelligence Research Institute.

Heckman, J., Lochner, L., & Todd, P. (2003). *Fifty years of Mincer earnings regressions* (Working paper). NBER WP 9732.

Hernandez, D., & Brown, T. B. (2020). Measuring the algorithmic efficiency of neural networks. arxiv:2005.04305.

Hibbs, D. A., & Olsson, O. (2004). Geography, biogeography, and why some countries are rich and others are poor. *Proceedings of the National Academy of Sciences,101*, 3715–3720.

Hidalgo, C. (2015). *Why information grows: The evolution of order from atoms to economies*. Basic Books.

Hilbert, M., & López, P. (2011). The world's technological capacity to store, communicate, and compute information. *Science,332*, 60–65.

Jones, C. I. (1995). R&D-based models of economic growth. *Journal of Political Economy,103*, 759–84.

Jones, C. I. (1999). Growth: With or without scale effects? *American Economic Review,89*(2), 139–144.
Jones, C. I. (2005). Growth and ideas. In P. Aghion & S. Durlauf (Eds.), *Handbook of economic growth*.
Jones, C. I. (2005). The shape of production functions and the direction of technical change. *Quarterly Journal of Economics,120*, 517–549.
Jones, C. I., & Kim, J. (2018). A Schumpeterian model of top income inequality. *Journal of Political Economy,126*, 1785–1826.
Jones, C. I., & Romer, P. M. (2010). The new Kaldor facts: Ideas, institutions, population, and human capital. *American Economic Journal: Macroeconomics,2*, 224–245.
Jumper, J., Evans, R., Pritzel, A., et al. (2021). Highly accurate protein structure prediction with alphafold. *Nature,596*, 583–589.
Kaldor, N. (1961). Capital accumulation and economic growth. In L. A. Lutz & D. C. Hague (Eds.), *The theory of capital* (pp. 177–222). Palgrave Macmillan.
Karabarbounis, L., & Neiman, B. (2014). The global decline of the labor share. *Quarterly Journal of Economics,129*(1), 61–103.
Klump, R., McAdam, P., & Willman, A. (2007). Factor substitution and factor augmenting technical progress in the US. *Review of Economics and Statistics,89*, 183–192.
Klump, R., McAdam, P., & Willman, A. (2012). Normalization in CES production functions: Theory and empirics. *Journal of Economic Surveys,26*, 769–799.
Kremer, M. (1993). Population growth and technological change: One million B.C. to 1990. *Quarterly Journal of Economics,108*, 681–716.
Kruse-Andersen, P. K. (2017). *Testing R&D-based endogenous growth models* (Working paper). University of Copenhagen.
Krusell, P., Ohanian, L. E., Ríos-Rull, J.-V., & Violante, G. L. (2000). Capital-skill complementarity and inequality: A macroeconomic analysis. *Econometrica,68*, 1029–1054.
Lucas, R. E. (1988). On the mechanics of economic development. *Journal of Monetary Economics,22*, 3–42.
Madsen, J. (2008). Semi-endogenous versus Schumpeterian growth models: Testing the knowledge production function using international data. *Journal of Economic Growth,13*, 1–26.
Meadows, D. H., Meadows, D. L., Randers, J., & Behrens, W. W. (1972). *The limits to growth*. Potomac Associates—Universe Books.
Meadows, D. H., Randers, J., & Meadows, D. L. (2004). *Limits to growth: The 30-year update*. Chelsea Green Publishing.
Mincer, J. (1974). *Schooling, experience, and earnings*. NBER Press.
Muehlhauser, L., & Salamon, A. (2012). Intelligence explosion: Evidence and import. In A. Eden, J. Søraker, J. H. Moor, & E. Steinhart (Eds.), *Singularity hypotheses: A scientific and philosophical assessment*. Springer.
Pacho, W. (2013). *Stagnacja gospodarcza i innowacje techniczne w okresie przedindustrialnym*. Oficyna Wydawnicza SGH.
Piketty, T., & Saez, E. (2003). Income inequality in the United States, 1913–1998. *Quarterly Journal of Economics,118*.
Piketty, T., & Zucman, G. (2014). Capital is back: Wealth-income ratios in rich countries, 1700–2010. *Quarterly Journal of Economics,129*(3), 1255–1310.
Raichle, M. E., & Gusnard, D. A. (2002). Appraising the brain's energy budget. *Proceedings of the National Academy of Sciences,99*(16), 10237–10239.
Romer, P. M. (1986). Increasing returns and long-run growth. *Journal of Political Economy,94*(5), 1002–1037.
Romer, P. M. (1990). Endogenous technological change. *Journal of Political Economy,98*, S71–S102.
Solow, R. M. (1956). A contribution to the theory of economic growth. *Quarterly Journal of Economics,70*, 65–94.
Tegmark, M. (2017). *Life 3.0: Being human in the age of artificial intelligence*. Knopf.

Uzawa, H. (1965). Optimum technical change in an aggregative model of economic growth. *International Economic Review,6*, 18–31.
Yudkowsky, E. (2013). *Intelligence explosion microeconomics* (Technical report 2013-1). Machine Intelligence Research Institute.

Chapter 5
Mechanization, Automation and the Labor Market

In the 21st century we might witness the creation of a massive new unworking class: people devoid of any economic, political or even artistic value, who contribute nothing to the prosperity, power and glory of society. This "useless class" will not merely be unemployed – it will be unemployable.

Yuval Noah Harari

5.1 Human Skills: Augmented or Replaced?

Before addressing the fundamental question about the future of humans and the human civilization at large, first I'd like to tackle the question about the future of human work. In this regard, the hardware software model of economic growth organizes the past evidence on long-run employment trends and offers interesting (though sometimes frightening) predictions for the future. The key dichotomy present in the model underscores that very different implications should be expected to follow whether machines come to replace human brawn (hardware) or brains (software). The model also clearly delineates few concepts that are almost always confused in economics discussions: mechanization, automation, adoption of information and communication technologies (ICT), development of AI, and robotization. Let me explain.

Mechanization of production consists in substituting human (and animal) physical labor with machines. While some degree of mechanization was observed since the early hominins developed their first stone tools, large-scale mechanization has been going on only since the industrial revolution. It applies to the physical actions (hardware) but not the instructions (software)—which are still provided by humans, according to their expertise and judgment. Therefore, mechanization replaces human physical work but *augments* human cognitive work. Augmentation means an increase

J. Growiec, *Accelerating Economic Growth*, Frontiers in Economic History,
https://doi.org/10.1007/978-3-031-07195-9_5

in the factor's marginal productivity, which typically translates into higher remuneration. An example of mechanization is when trains and automobiles replace human and horse muscles in performing the action of getting a person or a cargo shipment from place A to place B, or when indoor plumbing replaces the tedious chore of bringing clean and removing dirty water from the household.

In contrast, *automation* of production consists in substituting humans with pre-programmed software and AI algorithms in providing instructions (software) to machines (programmable hardware). Automation pertains to cases where a task, previously involving human cognitive work, judgment and decisions, is carried out entirely by machines without any human intervention (or perhaps with just a little human oversight and troubleshooting). Automation was very rare in history until information technologies came into use as general purpose technologies during the digital revolution. Routine tasks (both physical and cognitive) are the first to be automated (Bresnahan and Trajtenberg, 1995; Autor and Dorn, 2013; Cortes et al., 2017). An example of automation is when the former cognitive work of tax consultants is now carried out by tax calculating algorithms. And did you know that a *computer* used to be a job title?

Historically, automation came later than mechanization and was applied usually to tasks that had been already mechanized. In such cases, ultimately both human brawn and brains are replaced, leaving no human skills to be augmented.

As a curiosity one can also conjure cases of automation without mechanization. This is when a machine does the thinking and you do the action. In such a case, automation is a mirror image of mechanization: human cognitive work is replaced and physical work is augmented. It may happen, for example, when you walk around an unknown neighborhood blindly following the instructions of your GPS. (Hopefully, you were at least involved in choosing the destination.)

ICT adoption, in turn, refers to the simultaneous adoption of two complementary and mutually compatible production inputs, programmable hardware and pre-programmed software. The invention of programmable hardware constitutes a breakthrough in the history of technology because of its ability to store and run code which—in contrast to human brains—is non-rivalrous. Once programmed, it can be run an arbitrary number of times, possibly concurrently on many machines, without the need of any human intervention. For these reasons, ICT was necessary for initiating large-scale automation.

Further progress in automation requires the *development of AI*, that is, software that has the ability to learn from data. In contrast to "traditional", "non-intelligent" software, which consists of a deterministic, fixed set of instructions (e.g., if–then loops), artificial intelligence algorithms can improve their performance with experience and new information. Learning can happen both under a static and dynamic architecture of the AI. To this day, successful AI algorithms had fixed code and could only adjust parameters, but it is conceivable that in the future AI may also modify its own architecture while heading towards self-improvement.

Machine learning bears a lot of similarity to human learning: when provided with sufficient energy and relevant data, both processes generate visible increases in skill levels. The advantage of AI is, however, that many networked pieces of equipment can pool their data whereas humans can't. For example, an unusual road event encountered by a self-driving Tesla car can improve the driving skills of the entire fleet, whereas each human driver learns almost exclusively from her own experiences.

The development of AI opens new opportunities for speeding up automation because AI allows to substitute humans in a wider variety of tasks, including non-routine tasks as well. While computers drastically lowered the costs of computing (arithmetic), AI drastically lowers the costs of *prediction* (Brynjolfsson et al., 2019; Agrawal et al., 2017).

Finally, *robotization* is a process that works in parallel with automation but in the domain of hardware. It consists in replacing some of the human physical labor as well as the work of non-programmable machines with *robots*—multi-function autonomous devices that embody ICT and are able to work without human oversight. Industrial robots are abound in the motor and electronic industries. Their numbers are also on the rise in other industries and selected service sectors such as trade and logistics. We are also increasingly using service robots, in particular household robots such as autonomous vacuum cleaners or lawnmowers. According to the IFR (2017) definition, a multi-function industrial robot is "an automatically controlled, reprogrammable, multipurpose manipulator programmable in three or more axes, which may be either fixed in place or mobile for use in industrial automation applications". An analogous definition can be formulated for service robots.

In light of the above discussion, it is a misconception to identify computers and robots with their hardware. Computers and robots can be productive and useful as a whole only when the hardware is provided with appropriate instructions (code), i.e., software, either from human operators or pre-programmed software (or both). Another frequent misconception is to automatically associate AI with robots. AI is software that can learn from data. This software may indeed provide instructions to robots, but also to conventional computers and other machines which embody digital chips—and sometimes even to us humans (Benzell et al., 2015; Berg et al., 2018).

It is also rather problematic to identify AI development with automation (Aghion et al. 2019), because automation may proceed also without AI, as it has been the case for decades, e.g., in the auto industry, and AI doesn't have to have an immediate impact on productivity.

Crucially, however, automation must not be conflated with mechanization. This is probably the most common and most consequential mistake made in contemporary economics and policy discussions. As a telling example, look at the famous question: "will humans go the way of horses?" (Brynjolfsson and McAfee, 2014), i.e., will technological progress make humans as irrelevant for the economy as horses? For anyone acquainted with the hardware–software model, this is a very annoying question. As far as human physical labor is concerned, humans have long went the way of horses. But it hasn't been the case, or at least not yet, for cognitive tasks—for which horses are of no use.

Similarly, it is false comfort to say—and it's unfortunately a really common mantra—that "the history of the industrial revolution teaches us that when jobs are destroyed, new ones are bound to emerge". No. The industrial revolution only teaches us that when human physical labor is mechanized, additional workers will be demanded in cognitive occupations. It tells us nothing about cognitive occupations being automated.

So do new technologies augment or replace human skills? The hardware–software model sheds new light on this question by pointing out that hardware and software are mutually complementary whereas components within these two factors are substitutable. Therefore, mechanization—development of more powerful hardware—replaces human physical labor (another component of hardware) but augments software, including in particular human cognitive work. Mechanization was arguably the fundamental cause of the systematic upward wage trend observed since the inception of the industrial era. By contrast, automation—development of more powerful software—replaces human cognitive work (another component of software) but augments hardware. The pity is that after completing the process of mechanization, there is hardly any human contribution to hardware anymore. For this reason, automation (among possible other mechanisms) underpins the secular downward trend in the labor share observed globally since the 1980s (Karabarbounis and Neiman, 2014; Piketty and Zucman, 2014). Extrapolating past progress in automation into the future and assuming that AI will continue to improve its adaptivity and versatility, the hardware–software model delivers a prediction that automation will indeed give rise to a growing group of people who won't have any relevant skills suitable for the labor market. This is the group that Yuval Noah Harari called the "useless class" (Harari, 2017).

5.2 Will Humans Go the Way of Horses?

So—to repeat the annoyingly phrased but commonly asked questions—will humans become as irrelevant for the economy as horses? Will robots take all our jobs? Two influential articles, followed by a couple of corporate reports with a more international perspective, provided estimates of the probability that certain jobs will be automated within the next 20 years, based on present technology (Frey and Osborne, 2017; Arntz et al., 2016; McKinsey, 2017; PwC, 2018). The forerunning article by Carl B. Frey and Michael A. Osborne, first released in 2013, estimates that as many as 47% of workplaces in the USA exhibit a high (>70%) probability of being automated. In particular, I have a 43% probability being automated as an economist, though only 4% as a social scientist (and writing daily stock exchange briefs is already automated). Other reports have produced broadly similar numbers for other developed countries, whereas due to methodological differences, the study of Arntz et al. (2016) produced much lower numbers.

Exact probabilities are of course highly uncertain and subject to dispute, not only on methodological grounds but also due to other reasons ranging from the changing task content of jobs and innumerable institutional (e.g., legal, political) factors, to unpredictable future breakthroughs in AI. And I'm singling out future developments in AI here because they may really turn the tables upside down: AI has a long history of showing that seemingly difficult problems (e.g., playing chess) can turn out easy to automate, while seemingly easy problems (e.g., walking upright on varied terrain) can be astonishingly difficult.

The key take-away message from this literature is that we should expect many jobs to be automated in the coming decades, and particularly so routine jobs, both manual and cognitive, thus far performed by, e.g., telemarketers, cashiers, bank consultants, credit analysts, etc. In consequence, labor market polarization—the gradual disappearance of middle-class jobs—will probably be exacerbated in the future, and the least susceptible jobs beside AI researchers will require both advanced cognitive and social skills (e.g., pediatrists, primary school and pre-school teachers, psychotherapists, scientists).

But automation won't stop at routine occupations. Since the 1980s, growth in computing power and data storage has by far outstripped growth in global GDP per worker (Hilbert and López, 2011). Prospective returns to computer software have grown huge. Those who will manage to tap to the accumulated hardware and data overhang will therefore literally earn fortunes. The software champions of 2022, such as Google, Facebook, Microsoft, Amazon, Tencent or Alibaba, are at the forefront of a new wave of automation that will certainly continue into the future and likely accelerate.

My prediction that automation will likely accelerate is based on the expectation that in the future the "software" input to production will to an increasing degree include AI. By getting more versatile and adaptive, AI will replace human skilled labor in an increasing variety of tasks. It will be increasingly relevant also in non-routine tasks requiring advanced skills. Currently, automated image recognition is helpful in scanning X-ray images for signs of tumors (Autor and Dorn, 2013; Frey and Osborne, 2017; Berg et al., 2018)—but this is an exception rather than a rule. However, sufficiently improved AI algorithms will make it possible to automate also highly skilled jobs, including the ones that require highly specialized skills and offer competitive salaries. According to a poll of machine learning researchers (Grace et al., 2017): "AI will outperform humans in (...) translating languages (by 2024), writing high-school essays (by 2026), driving a truck (by 2027), working in retail (by 2031), writing a bestselling book (by 2049), and working as a surgeon (by 2053). Researchers believe there is a 50% chance of AI outperforming humans in all tasks in 45 years and of automating all human jobs in 120 years".

Such point estimates are, of course, clouded by huge uncertainty, and therefore should be taken with a grain of salt. AI researchers—and the general public alike—span the entire spectrum, from techno-enthusiasts like Ray Kurzweil who firmly predicts a technological "singularity" around 2045 (Kurzweil, 2005), to scholars who equally firmly believe that artificial general intelligence (AGI) will never be constructed and automation of all human work will never happen. Specialists frequently emphasize that even though progress in AI has accelerated in the last decades, achieving further progress remains hard. So hard that, as another poll of AI researchers states, "according to 92.5% of the respondents, superintelligence is beyond the foreseeable horizon" (Etzioni, 2016). But then again, can we really trust the time-frame of a "foreseeable horizon" given all the surrounding uncertainty and well-documented cognitive biases in human thought?

5.3 Complex Tasks and Economic Linkages

While the terminal point of automation is likely an economy that doesn't need human skills anymore, interesting things are also bound to occur on the run-up. A key factor that weighs in along the transition to a fully automated economy is growing task complexity.

Have you ever considered why the best jobs of today—high-salary, low-routine occupations that are least susceptible to automation—require both advanced cognitive and social skills, and a fair amount of responsibility? Precisely because they include tasks that are highly *complex*. Such tasks consist of multiple consecutive and parallel steps that must be performed in unstructured, uncertain environments, and often in cooperation with many other people. A complex task is successfully finished only after *all* of its multiple components (sub-tasks) are finished, which means that failure of a single sub-task may compromise the execution of the entire task. (Just think about managing a construction site, designing an IT solution for a large business enterprise, or flying a passenger jet. There's so many things to take care of, and none of them is allowed to go terribly wrong. See Kremer (1993); Growiec (2013).) Task complexity adds an additional layer of complementarity within the software factor: within sub-tasks, human and machine cognitive skills remain substitutable, but as the sub-tasks are complementary to one another, the overall degree of their substitutability or complementarity may vary (Growiec, 2022a).

Growing task complexity creates thrilling job opportunities. Since the dawn of the digital era, it is getting increasingly valuable to engage in tasks, which include both sub-tasks that are automated and ones for which automation is not yet possible. Such jobs ride on the temporary complementarity between human and machine cognitive work within complex tasks. For example, pre-programmed computation routines are used in book-keeping, inventory management, airplane piloting, data analysis and forecasting, engineering and architecture, and cutting-edge science. In all such cases, algorithms are used to automate the more tedious and repetitive parts of the cognitive process, vastly augmenting our human skills and allowing our brains to stay focused on the less structured parts that involve decision-making under uncertainty and imperfect information. One of the top career advices I've heard, by Hal Varian, Google Chief Economist, is: "seek to be a scarce complement to increasingly abundant inputs".

Sub-task complementarity stretches also into the domain of more advanced algorithms, including AI. Fruitful *human–AI cooperation* is already happening. For example, my writing of this book has been greatly facilitated by the Google search engine which helped me locate literature sources dealing with various problems that I was uncertain about. In arts, Taryn Southern's 2018 pop music album "I am AI" was created in collaboration with Amper—an AI music composition software. In business, increasing shares of revenue are generated thanks to online ads that are tailored to the viewers' preferences revealed by AI algorithms. In finance, a vast majority of stock trades are generated by high-frequency trading software. In science, machine learning is applied in genome sequencing and identification of astronomical objects. The general pattern is that humans are gradually specializing in sub-tasks that involve high-level judgment and decision-making, whereas pre-programmed software does the repetitive computations and AI excels in prediction (Agrawal et al., 2017, 2019). A concurrent trend is that job descriptions are getting more and more complex.

Historically, computer algorithms have always been a subordinate of the human, a tool in our hands and for our heads. They have augmented our cognitive skills, empowered us to transcend our limits, and allowed us to claim the residual value. It is unclear, however, if this state of affairs will last forever. There is a possibility that the sweeping progress of AI-fueled automation may eventually catch up with human capabilities at breadth, and more and more tasks will be *fully* automated, leaving no value to be added by human work.

In this regard, Acemoglu and Restrepo (2018) have a model of "a race between man and machine", or better: between human cognitive work and AI. It describes a world where all existing tasks are gradually automated, but on top of that people continuously create new tasks in which they maintain an edge over AI. Under a certain parametrization, the model economy converges to a long-run equilibrium where the share of automated tasks is constant and there is always some cognitive work for us to do. However, a key assumption of this model is that people will always be initially better at performing the newly created tasks. I worry that this assumption may one day be invalidated. (And even before it is invalidated, there may be only few people on the planet able to stay ahead of automation. The rest would hopelessly observe their skills become irrelevant for the economy.) It may turn out that from a certain point in time onwards, newly created tasks will be fully automated from the outset, with no sub-tasks left for humans to perform. After all, biggest gains from automation accrue when tasks are fully automated, don't they?

Complexity of tasks relates closely to the structure of economic linkages. Modern technologies embody knowledge from a variety of different disciplines, and produced goods embody parts from different sectors of the economy. Decades of incremental technological progress have made our products into intricate assemblies of a large number of intermediate goods coming from all economic sectors and from all over the world. Concurrently, decades of process innovations have created a range of auxiliary operations—such as business-to-business services—that are necessary to perform the central tasks efficiently enough to stay in the market. However, this network of linkages can be fundamentally disrupted when major tasks are fully automated. For example, the arrival of Amazon has profoundly shaken the book-selling business across the world, leading many retailers to bankruptcy and forcing others to change the scope of their activity. Automated processes use inputs in different proportions, and computers have different needs than people. This changes the input–output matrix and modifies supply chains, making equilibrium consequences of task automation hard to predict (Brynjolfsson et al., 2019).

Nevertheless, despite all this complexity some first-order effects can still be assessed. First, jobs will be lost due to automation. Second, jobs that are complementary to the automated tasks will temporarily rise in value, until they will be automated, too. Third, ongoing technological progress will create a stream of new, increasingly complex tasks and jobs. In at least some of them human brains should have an initial advantage over computer algorithms. In the long run, though, there are reasons to be anxious about the outcome of the "race between man and machine".

The fundamental question of whether *all* jobs will be eventually automated, is ultimately equivalent to the question whether AI will one day reach superhuman intelligence. (Without that, *some* jobs should remain safe.) We don't know this, but we certainly know that there are no physical constraints that would make it impossible. As Stephen Hawking once said, and I fully endorse, "there is no physical law precluding particles from being organized in ways that perform even more advanced computations than the arrangements of particles in human brains". Therefore, while in the short run, certain human skills may be complementary to AI, we should prepare that, in the long run, the labor market may gradually disappear. All our skills may in the long run be replaced by more intelligent, efficient, reliable computer algorithms. Nevertheless, this "long run" is probably not yet around the corner.

5.4 Automating Research

Thus far, GDP growth doesn't catch up with growth in the digital sphere because human cognitive work, whose global supply grows much less rapidly, remains an essential production factor. This scarcity can be alleviated only through full automation, unpinning the volume of instructions provided to hardware from the size of the human population. But as long as automation is only partial, a key growth engine is the production of non-rivalrous technological ideas which increase human productivity. Automation of R&D activities, in turn, constitutes an interaction of both growth engines, automation and innovation. Therefore, if only possible to achieve, it has the potential to create a virtuous developmental circle that substantially accelerates growth (Growiec, 2022b).

Automation in R&D is certainly possible and to an extent already happening. The preliminary developments in AI that we've seen so far have already shaken the stereotypical belief that science is too complicated to be ever automated. To paraphrase the words of Elon Musk, humans are generally dumber than we think we are, and this applies particularly to those of us who have built their identity around "being smart": professors, researchers, authors, entrepreneurs, etc. The pattern of systematic improvement in capabilities of AI algorithms reshapes our understanding of what is easy and what is hard. All too often it turns out that what feels easy to humans—say, a leisurely stroll on a grass meadow or empathizing with a crying child—can be actually very hard from the perspective of information requirements and algorithmic complexity. On the other end of the spectrum, by contrast, some of the most spectacular landmarks of "being smart" have already fallen to AI, like chess, when the ruling world champion Garri Kasparov was defeated by Deep Blue in 1997, *Jeopardy!*, when quiz show prodigies Ken Jennings and Brad Rutter lost to IBM Watson in 2011, and Go, as the ruling world champion Lee Sedol was defeated by DeepMind AlphaGo in 2016. (Since these historical moments, the advantage of AI over humans in these three fields only skyrocketed.)

In science AI is used, for example, in genome sequencing, analysis of astronomical data, and automatic generation of decision rules from big data sets. It is also applied to enhance drug discovery (Agrawal et al., 2019), predict protein structures (Jumper et al., 2021), and help mathematicians prove theorems (Davies et al., 2021). And even though R&D tasks have been relatively resilient to automation thus far because of their complex, non-routine character, it's still fair to say that most recent technological developments would have never been obtained if not for programmable *R&D capital*: sophisticated, computerized physical, chemical and pharmaceutical laboratories and the abundance of general-purpose computing power.

In the digital era, computational complexity of our research grows hand in hand with computing power, allowing to build more sophisticated models of reality, to pursue more detailed empirical identification of facts based on statistical data, and to engage in extensive exploratory research and data mining. Digital era technologies also allow to instantly search the massive and exponentially growing base of earlier publications.

As a researcher, and indeed a person who builds his identity around "being smart", I find this observation quite humbling. Extrapolating the past progress, over the coming decades I expect a sweep of progress in sciences, humanities, design, and even arts, driven by automation, big data analysis and AI. This will make the work of human researchers, and my own work in particular, increasingly irrelevant and substandard—unless we manage to learn not to just use specialized computer tools, but to actually overcome our mounting cognitive bandwidth problem and engage in productive cooperation with AI.

The perspectives of automating R&D have such a large bearing on future growth also because more and more resources are allocated to R&D these days. The ratio of global R&D spending to GDP is higher than ever before, and it's bound to increase even further. Research and innovation policies are part and parcel of all contemporary economic development policies. Just as "scalability" is buzzword no. 1 in Silicon Valley, "innovativeness" is buzzword no. 1 in economic policy circles. Across the world, the R&D share of output is also positively correlated with output itself. Apart from a few exceptions like oil-producing countries of the Persian Gulf, the richest countries are also the most innovative.

Extrapolating current trends into the future, and keeping in mind that R&D is a source of economic increasing returns because information is non-rivalrous, I expect rapid knowledge accumulation with an ever larger fraction of inputs assigned to the R&D process, and an increasing contribution to its output coming from R&D capital as opposed to R&D labor. This may not be the end of the story, though, because progress in AI capabilities is nonlinear (Brynjolfsson and McAfee, 2014; Acemoglu and Restrepo, 2018; Frey and Osborne, 2017). Advanced AI algorithms may one day not merely complement human cognitive skills in carrying out the more mundane, routine tasks, but substitute out humans altogether, gradually conquering at all stages of the R&D process, from basic research to practical implementation.

The discussion here goes back once again to the question of complementarity versus substitutability between pre-programmed software (inclusive of AI algorithms) and human cognitive work. Thus far R&D has been one of the strongholds of human thought, with humans making all the important decisions and software being only a tool in researchers' hands. The primary reason for this apparent complementarity is that computer algorithms have been historically very bad at ideation, creativity and asking useful research questions (Brynjolfsson and McAfee, 2014), so that certain necessary steps of the R&D process were not automatable. But this doesn't have to be always the case. Day by day, AI algorithms are getting better and better at pattern recognition based on big data, classification, categorization of various sorts of content, and making adaptive decisions in noisy, variable environments—and they are much faster than humans at all that. In the future, highly developed machine learning and big data algorithms, autonomous laboratories, automatic translators, text generators, and multi-function robots, may allow multi-step R&D processes to be *fully* automated. And then the fundamental substitutability between human cognitive work and AI, predicted by the hardware–software model, will kick in.

Brynjolfsson and McAfee (2014) predict that sophisticated AI technologies will likely develop "gradually, then suddenly", fueled by their highly scalable character and—potentially—ability to self-improve. (Just like in the famous quote of Ernest Hemingway: "How did you go bankrupt? Two ways. Gradually, then suddenly.") I totally agree with this prediction. In the coming decades economic growth at the global technological frontier will be driven by automation, innovation and—to an increasing extent—automated innovation.

5.5 Human Control and the Useless Class

How many of us will be able to cooperate with AI in performing increasingly complex tasks, and how many of us will be excluded? Automation is already affecting our labor markets, and along with its future advances, it will certainly bring even more profound implications for employment and economic inequality. The famous historian/futurist Yuval Noah Harari expects that further progress in automation and AI will inevitably create a growing *useless class* in future societies (Harari, 2017).

Looking at the labor markets of 2022, one may be tempted to dismiss this claim. Unemployment rates are low, wages are growing, educational attainment and labor market participation (particularly of women) are at their historical high. Isn't it that even if some jobs in the industry are automated, new jobs are created elsewhere, for example in services, and people always find their way to fill them? Well, so far, so good, but when thinking of the long run, there is a good reason to agree with Harari, at least partly.

Namely, AI capabilities are advancing fast—and certainly way faster than our human skills which are (or otherwise soon will be) reaching a plateau (Barro and Lee, 2013). And when jobs are automated, people have only two ways to go: up or out. The only chance to remain employed is to move towards jobs that are more *difficult for AI*. (Note that this doesn't necessarily mean towards high-skilled jobs; to the contrary, people have an advantage over AI also in jobs that are traditionally considered low-skilled, such as for example jobs requiring personal care, empathy, or social interaction, e.g., nurse, babysitter, secretary, etc.). But the tide is rising and there will certainly be more and more people without *any* automation-proof skills.

However, not all which can be automated, will be automated. Automation technologies, just like human work, have a price, so it could turn out that human work would still be cheaper. Specifically, if people would be ready to work for any wage, we wouldn't be worrying about the threat of technological unemployment, but rather about stagnant wages amidst ongoing economic growth (Growiec, 2022a). People won't work for any wage, though, unless they really like their work and have a secure non-labor source of income. In sum, I believe that Harari's claim about the useless class is correct but subject to an important qualification. Emergence of a useless class is indeed likely, but it is predicated on the assumption that the affected people won't be willing to work for the equilibrium wage accruing in their automatable profession. That wage will likely stagnate, and surely won't be able to keep up with GDP, implying that its share in aggregate output will decline. (The reservation wage, i.e., minimum wage at which people are willing to work, will also be affected by their non-labor income, including capital income and social transfers.)

Meanwhile, the spread of AI applications will change our perception of who's smart and who's not. For example, the genius mathematician John von Neumann stunned his contemporaries in how fast and accurately he calculated, to the point that Nobel Laureate, physicist Hans Bethe said "I have sometimes wondered whether a brain like von Neumann's does not indicate a species superior to that of man". To an AI, however, John von Neumann's analytical brain may seem all too human, perhaps *human beyond repair*: what's easier than a few algebraic manipulations? (We've already seen this shift of perspective in the case of chess, Go, or *Jeopardy!*) In the future, even highly advanced analytical skills may be more automatable than, say, certain artistic or social skills (Yudkowsky, 2013).

For this reason, in economically advanced countries like the US automation is already causing income polarization, with a disappearing middle class that used to perform routine jobs, like assembly workers or office clerks. People who will find themselves unable to switch to non-routine jobs will likely be the first to become entirely useless for the labor market. Stereotypically, the hardest hit should probably be narrow-minded specialists with little adaptation potential, low social skills and empathy, and anxious of job change and migration. Specifying the demographics of the prospective excluded group, the *useless class* (Harari, 2017), is, however, highly uncertain because we don't know the order in which jobs will be automated, the pace of the process and its spread across the globe. We also don't know how big people's ability is to adapt to the new environment and "reinvent themselves" for a new job, especially if it's going to be a less respectable one, less paying one, or one in a different part of the world. Nevertheless, a relatively safe speculation is that the future useless class will be cross-cutting the current societal divides, including individuals both from groups that are already subject to exclusion and groups that now maintain a high status. Junior executives and specialists may be fired in automation-driven corporate reorganization events earlier than the company's cleaning staff. But even then it's unclear how things will pan out in the end: perhaps some of the displaced specialists will be later hired as the new cleaning staff, pushing the former ones into permanent unemployment?

To assess the possible social consequences of a rising useless class, one needs to address the following questions. First, do we really *need* to work? Is it necessary for stable development of the human civilization that a vast majority of adult humans have jobs? Second, do we *like* to work? Is some job always better than no job? These are tough questions that would have to be addressed rather soon. I am clearly not in the position to provide clear answers to them. Nevertheless, I would like to make two tangential comments.

First, be warned that the emergence of the useless class is going to be a global phenomenon and therefore should be best addressed globally. National policies for handling technological unemployment, like country-level Universal Basic Income schemes that are sometimes proposed, will only exacerbate international tension (Harari, 2017) and may be unsustainable.

Second, policies and incentives must be put in place so that people don't feel useless *as a whole* because *their work* has become redundant. We are control-maximizing animals and we love the feeling of being able to choose and exert control over our lives and our surroundings. And if the only admissible solution is to provide us with an illusion of control, it better be a convincing illusion. Indeed, even though thus far employment has been one of the key sources of life satisfaction, and our occupations were what defined us as people (as in: "Who are you? An economist/lawyer/teacher/project manager/etc."), this is not automatically guaranteed to persist into the future. It may have arisen as a reflection of human work being the cornerstone of the economy. In the age of advanced automation, human inputs will gradually cease to be central for production, though. Then, if the world finds a way to distribute the wealth widely (a big "if"...), the solution must be to find ourselves other ways than work to pursue local control and satisfy our hardwired needs of social belonging, esteem and self-actualization. Perhaps, for example, through digital creativity?

References

Acemoglu, D., & Restrepo, P. (2018). The race between man and machine: Implications of technology for growth, factor shares and employment. *American Economic Review,108*, 1488–1542.

Aghion, P., Jones, B. F., & Jones, C. I. (2019). Artificial intelligence and economic growth. In A. Agrawal, J. S. Gans & A. Goldfarb (Eds.), *The economics of artificial intelligence: An agenda*. University of Chicago Press.

Agrawal, A., Gans, J. S., & Goldfarb, A. (2017). What to expect from artificial intelligence. *MIT Sloan Management Review*, Spring.

Agrawal, A., Gans, J. S. & Goldfarb, A. (2019). Artificial intelligence: The ambiguous labor market impact of automating prediction. *Journal of Economic Perspectives,33*, 31–50.

Arntz, M., Gregory, T., & Zierahn, U. (2016). *The risk of automation for jobs in OECD countries: A comparative analysis, OECD social, employment and migration* (Working Paper No 189). OECD Publishing.

Autor, D. H., & Dorn, D. (2013). The growth of low-skill service jobs and the polarization of the US labor market. *American Economic Review,103*(5), 1553–97.

Barro, R., & Lee, J.-W. (2013). A new data set of educational attainment in the world, 1950–2010. *Journal of Development Economics,104*, 184–198.

Benzell, S. G., Kotlikoff, L. J., LaGarda, G., & Sachs, J. D. (2015). *Robots are us: Some economics of human replacement* (Working Paper No. 20941). NBER.

Berg, A., Buffie, E. F., & Zanna, L.- F. (2018). *Should we fear the robot revolution? (The Correct Answer is Yes)* (Working Paper WP/18/116). IMF.

Bresnahan, T. F., & Trajtenberg, M. (1995). General purpose technologies: Engines of growth? *Journal of Econometrics,65*, 83–108.

Brynjolfsson, E., & McAfee, A. (2014). *The second machine age: Work, progress, and prosperity in a time of brilliant technologies*. Norton & Co.

Brynjolfsson, E., Rock, D., & Syverson, C. (2019). Artificial intelligence and the modern productivity paradox: A clash of expectations and statistics. In A. Agrawal, J. S. Gans & A. Goldfarb (Eds.), *The economics of artificial intelligence: An agenda*. University of Chicago Press.

Cortes, G. M., Jaimovich, N., & Siu, H. E. (2017). Disappearing routine jobs: Who, how, and why? *Journal of Monetary Economics,91*, 69–87.

Davies, A., Veličković, P., Buesing, L., et al. (2021). Advancing mathematics by guiding human intuition with AI. *Nature,600*, 70–74.

Etzioni, O. (2016). No, the experts don't think superintelligent AI is a threat to humanity. *MIT Technology Review*

Frey, C. B., & Osborne, M. (2017). The future of employment: How susceptible are jobs to computerisation? *Technological Forecasting and Social Change,114*, 254–280.

Grace, K., Salvatier, J., Dafoe, A., Zhang, B., & Evans, O. (2017). *When will AI exceed human performance? evidence from AI experts* (Working paper). Future of Humanity Institute, Oxford University.

Growiec, J. (2013). A microfoundation for normalized CES production functions with factor-augmenting technical change. *Journal of Economic Dynamics and Control,37*, 2336–2350.

Growiec, J. (2022a). Automation, partial and full. *Macroeconomic Dynamics,26*.

Growiec, J. (2022b). What will drive global economic growth in the digital age?. *Studies in Nonlinear Dynamics and Econometrics*, forthcoming.

Harari, Y. N. (2017). *Homo deus: A brief history of tomorrow*. Vintage.

Hilbert, M., & López, P. (2011). The world's technological capacity to store, communicate, and compute information. *Science,332*, 60–65.

IFR. (2017). *The impact of robots on productivity, employment and jobs*. International Federation of Robotics, Positioning paper.

Jumper, J., Evans, R., & Pritzel, A., et al. (2021). Highly accurate protein structure prediction with alphafold. *Nature,596*, 583–589.

Karabarbounis, L., & Neiman, B. (2014). The global decline of the labor share. *Quarterly Journal of Economics, 129*(1), 61–103.

Kremer, M. (1993). The O-ring theory of economic development. *Quarterly Journal of Economics, 108*, 551–575.

Kurzweil, R. (2005). *The singularity is near*. Penguin.

McKinsey. (2017). *A future that works: Automation, employment and productivity*. McKinsey Global Institute: Report.

Piketty, T., & Zucman, G. (2014). Capital is back: Wealth-income ratios in rich countries, 1700–2010. *Quarterly Journal of Economics, 129*(3), 1255–1310.

PwC. (2018). *Will robots steal our jobs? an international analysis of the potential long-term impact of automation*. PriceWaterhouseCoopers: Report.

Yudkowsky, E. (2013). *Intelligence explosion microeconomics* (Technical report 2013-1). Machine Intelligence Research Institute.

Chapter 6
Scale of Operations

The melting of the polar ice caps at the end of the ice age, around 10,000 B.C., and the consequent flooding of land bridges provide a natural experiment that nearly eliminated contact between the old world, the Americas, mainland Australia, Tasmania and Flinders Island. (...) In 1500, just after Columbus' voyage reestablished technological contact, the region with the greatest land area, the Old World, had the highest technological level. The Americas followed, with the agriculture, cities, and elaborate calendars of the Aztec and Mayan civilizations. Mainland Australia was third, with a population of hunters and gatherers. Tasmania, an island slightly smaller than Ireland, lacked even such mainland Australian technologies as the boomerang, fire-making, the spear-thrower, polished stone tools, stone tools with handles, and bone tools, such as needles. Flinders Island, near Tasmania, has only about 680 square kilometers of land, and according to radiocarbon evidence, its last inhabitants died out about 4000 years after they were cut off by the rising seas, suggesting possible technological regress.

Kremer (1993), referring to the work of Jared Diamond

6.1 Increasing Returns to Scale

An important factor that contributed to the long-run success of the human species in furthering its local control, conquering the world and building a global economy, are *increasing returns to scale* (IRS) in production and research and development (R&D). IRS mean that proportional increases in inputs yield more than proportional increases in output (Jones, 2005a). In consequence—in the absence of constraints or counterveiling forces—they allow economic development to perpetuate itself in a self-sustaining accelerating trend.

J. Growiec, *Accelerating Economic Growth*, Frontiers in Economic History,
https://doi.org/10.1007/978-3-031-07195-9_6

But why would returns to scale be increasing? After all, textbook treatments of the macroeconomic production function typically assume constant returns, so that proportional increases in inputs yield proportional increases in output. This is justified by the following *replication argument*. Assume there is a factory that uses x units of capital, employs y workers and produces z units of output a day. What would happen if you built a second, identical factory nearby, also employing x units of capital and y workers? Total production should then be $2z$ a day, shouldn't it?

Validity of the replication argument rests on two hidden assumptions, though. First, technology must be the same across all firm sizes. If instead the set of available technologies expanded with increased scale of production, then doubling the inputs would grant access to a more productive production technique—so that output would be more than doubled, implying IRS. Second, all inputs (such as capital and human work) must be rivalrous. Adding non-rivalrous inputs to the mix, like ideas or computer code, would also lead to IRS. Think, for example, about one computer equipped with x bytes of code that produces z bytes of useful information. What would happen if you had two such computers and $2x$ bytes of code? Well, since code is non-rivalrous, you could costlessly copy and simultaneously run the $2x$ bytes of code on both computers—and consequently produce more than $2z$ bytes of output.

A Nobel-prize winning observation by Paul Romer is that since technological ideas are non-rivalrous, they are a source of increasing returns to scale and thus an engine of growth over the long run (Romer, 1990). If under a fixed technology, doubling of rivalrous inputs such as capital or human work doubles the output, then what would happen if the technology level also doubled on top of that? Output would then clearly more than double—and that constitutes IRS. Therefore, Paul Romer says, if there are constant returns to scale with capital and labor (the rivalrous inputs), there will be increasing returns with capital, labor, and technology.

To appreciate Paul Romer's insight to its fullest, let me rephrase it in terms of the hardware–software model. Recall that hardware includes physical capital and human physical labor (both of which are rivalrous), whereas software includes human cognitive work (rivalrous) augmented by technology (non-rivalrous), as well as pre-programmed software and AI (non-rivalrous, too). What would then be the implications of doubling all hardware and software inputs? The answer is that output will more than double: some of the software inputs are non-rivalrous, and therefore a source of increasing returns to scale. Specifically, there are two distinct channels at work, the former present throughout the whole human history and the latter appearing only after the digital revolution. First, human cognitive work goes up either by (i) increasing employment in cognitive occupations or average employee skills, or by (ii) raising the level of available technology. Both factors are multiplicative and the latter is non-rivalrous. This means that doubling each of them at the same time actually delivers a fourfold overall software increase. Second, in the digital era, we also have a non-human software component, which may grow either when (i) the stock programmable hardware or the AI skill level (algorithmic efficiency) grows, or (ii) the level of available technology goes up. Again, both factors are multiplicative and the latter one is non-rivalrous, so when both factors double, the total software increase is fourfold.

In sum, technological progress—which adds to software accumulation—has been the key driver of economic growth throughout the centuries because technology is non-rivalrous and therefore a source of IRS. On top of that, in the digital era, accumulation of programmable hardware has double duty as it adds both to hardware and software. The replication argument breaks down over the long run precisely because of these two facts.

Non-rivalry of knowledge makes it very distinct from rivalrous capital and labor inputs: knowledge is cumulative, so that previous ideas act as broad basis for new ones, facilitating new discoveries and generating IRS. In contrast, accumulating physical capital doesn't facilitate its further accumulation: due to the law of decreasing marginal capital productivity—stemming from the fundamental complementarity of hardware and software—each additional unit of capital makes its further accumulation only more difficult. The same argument follows for human labor. Programmable hardware, in turn, is still a different animal. It has the potential to overcome the law of decreasing marginal productivity and become an additional source of IRS because it comes in a package with non-rivalrous pre-programmed software that is able to exploit its capacity without the need to employ rivalrous human cognitive work.

In the following sections, I will argue that the presence of increasing returns to scale in production and R&D was an important reason why the human local control maximization process was able to get out of evolution's hand—that is, why our species managed to conquer the world, produce a sequence of technological revolutions, and enter a path of accelerating growth. I will foremostly concentrate on software rather than hardware because that's where the increasing returns are located. Then I will discuss the gradual economic transition of humankind from being a collection of localized hunter-gatherer tribes, through the stages of much larger but still isolated farming communities, emerging trade linkages, fiercely fighting kingdoms and empires, independent nation states, to a globalized, digitally connected world. (Tangentially, see West (2017) for a broad overview of scaling phenomena in nature and the economy.)

Next I will critically address a popular view that institutions, and in particular state capacity, were the key trigger that set sustained economic growth in motion. Instead, in my perception, the causal link usually ran in the opposite direction throughout the human history: technology was first and institutions second. At the global scale, institutions were adapted to rather than had a decisive impact on the current technological and economic realities. The only cases where the vector of causality was reversed were when the policy was disastrous, for example during the early Middle Ages (sometimes referred to as Dark Ages), during the two world wars of the twentieth century, in Robert Mugabe's Zimbabwe or North Korea under the Kims. As Yuval Noah Harari put it, "politicians have a very limited capacity to do good, but an unlimited capacity to do harm". In other words, in my view, institutions are not the reason why the *world economy grows*, but only the reason why some *nations fail* (Acemoglu and Robinson, 2012; Harari, 2014). In this regard, I will argue that up till now, our policy is visibly lagging behind technology and the economy, and that global governance should be strengthened or otherwise we may—sooner rather than later—again see bad politicians doing major harm. (Alas, this warning has materialized already. As this book goes out to print in April 2022, we are witnessing the horrifying Russian military aggression on Ukraine, almost single-handedly decided by Russia's President Vladimir Putin. This provides the global community with a strong incentive to strengthen global governance, needed to stop the war as soon as possible, reinstate Ukraine's control over its territory, and make sure that such appalling events are successfully deterred in the future.)

6.2 Technological Accelerations

There were three major accelerations in the pace of software accumulation, the source of IRS, in the human history: the cognitive revolution (about 70 000 BP), the scientific revolution (ca. 1550 CE) and the digital revolution (ca. 1980 CE). Each of them brought qualitative changes to the processing and transmission of information across the human civilization.

The cognitive revolution of the Upper Paleolithic endowed the *homo sapiens* with important cognitive skills such as improved language, an advanced theory of mind and the ability to create, document and communicate gossip, fictional stories and abstract ideas (Dunbar, 1998; Harari, 2014). The acquired ability to join forces and spread ideas constituted a major leap compared to other species. The *homo sapiens* was the first species to pass the threshold of a cognitive steady state, i.e., long-run stagnation in knowledge, and enter a path of accelerating growth. As inferred from the variety of objects found in excavations, the humankind's stock of knowledge began to be generally trending up soon after the cognitive revolution. Still, in the hunter-gatherer era there was no conscious "research" activity and thus discoveries could only be obtained via learning by doing, chance experiments, and informal thinking. Folk wisdom was transferred orally across generations, often in religious or superstitious frames, and generally was not shared beyond the tribe. As knowledge was stored only in people's memory, it was often forgotten. Thus—in economic terms—the hunter-gatherer era knowledge accumulation process was characterized by relatively fast depreciation of knowledge, low efficiency of knowledge creation and small scale of operations (Diamond, 1997; Pacho, 2013). All these caveats notwithstanding, scaling up the size of the collective human brain from a single isolated individual to a band of 30–50 or a tribe of about 150 people (Dunbar, 1998), with a possibility of occasional inter-tribe information exchanges, was a tremendous leap.

The next "software" revolution was the scientific revolution initiated in sixteeth-century Europe. It's worth reflecting why the scientific revolution didn't happen centuries earlier. After all, in its core, it was based on ancient ideas like the alphabet, writing, arithmetic or philosophy, wasn't it? In particular writing, when exercised on durable material, could make knowledge essentially eternal, drastically reduce the risk that useful ideas get forgotten over generations, and reduce duplication in "research" activity. (Useful knowledge was still sometimes forgotten and had to be reinvented from scratch, though, as exemplified by the case of Roman bridges and aqueducts.) Yet, all these brilliant ideas turned out insufficient to trigger a scientific revolution in antiquity or the Middle Ages. Ancient Greeks and Romans achieved remarkable developments in science, philosophy and mathematics that are taught at school to this day but encountered limitations due to the relatively small scale of their operations and low efficiency of information transmission: the contemporary writing technology employed scribes, and so it was very time-consuming and costly to copy books and scripts. What was badly needed was a dissemination technology such as Johannes Gutenberg's printing press which would further expand the size of the collective human brain able to contribute its new ideas. Unfortunately, the printing press hasn't been broadly adopted until the fifteenth century CE. But when it finally arrived, it was followed by a series of breakthrough ideas which facilitated further discoveries and inventions, and hence increased the efficiency of knowledge creation: the university, research laboratory, and the modern scientific method—empiricism, experimentation, mathematization and hypothesis testing (Mokyr, 2016). In result, technological progress accelerated and increasing returns kicked in. Soon Europe was ready to initiate the industrial revolution and put economic growth in a higher gear once again.

The digital revolution initiated in the 1980s provided further acceleration in the pace of information transmission and processing. For the first time in history, information transmission was decoupled from the human ability to exchange knowledge (either in speech or through writing and reading), and information processing—from human cognition. A tremendous leap in this regard was obtained thanks to programmable hardware equipped with digital computing power, memory, and (since 2000s) broadband Internet connection. Apart from being an achievement on its own, the digital revolution was also instrumental in boosting the efficiency of knowledge creation: automation of tedious research tasks, digitalization of scientific knowledge, and the World Wide Web equipped with efficient search engines gradually improved our ability to distinguish facts from myths, organize our knowledge set, locate gaps in it, ask novel, meaningful research questions, formulate testable theories, collect relevant data, and pursue empirical verification of our theories. The same developments were also paramount in easing global dissemination of new knowledge and technologies. International technology diffusion has accelerated markedly compared to the earlier centuries: it now takes a few years for top-notch technologies like smartphones to be adopted worldwide, while for the steam engine and the steam locomotive it took many decades (Comin and Hobijn, 2010), not to mention that numerous regions of the world actually never built railroads. The digital economy is perhaps the best example ever of increasing returns to scale in production. *Scalability* is one of key buzzwords in the Silicon Valley.

Looking at the grand time scale, the accelerating progress in information transmission, processing and storage is best summarized by Gillings et al. (2016): "After RNA genomes were replaced with DNA, it then took a billion years for eukaryotes to appear, and roughly another two billion for multicellular organisms with a nervous system. It then took another 500 million years to develop neural systems capable of forming languages. From there, it took only 100,000 years to develop written language, and a further 4,500 years before the invention of printing presses capable of rapid replication of this written information. The digitalization of the entire stockpile of technologically-mediated information has taken less than 30 years. Less than 1% of information was in digital format in the mid-1980s, growing to more than 99% today".

6.3 Making of the Global Research Network

But what made these consecutive accelerations possible? Why didn't we at some point just stagnate in knowledge, as all other species do before competitors push them out of their ecological niches into extinction? Why have we passed this filter?

Well, this was neither automatic nor guaranteed; many things could go wrong along the way. But one factor stands out as an important catalyst of technological progress and economic growth: systematic increases in the scale of operations in R&D, allowing to exploit increasing returns to scale in the accumulation of software.

What I mean by *scale of operations* in this context is the degree of pooling of information and talent in the research and development (R&D) process. Increasing the scale of operations in R&D is paramount in accelerating growth because information—as opposed to physical goods—is non-rivalrous. Furthermore, knowledge accumulation is a cooperative process that occurs in a social network that makes up the *immortal collective brain* (see Hidalgo (2015) for a brilliant exposition of the network character of information accumulation). Even if some researchers may be working alone (and they were very often working alone in the past), their efforts become successful only after the new information is incorporated into the existing knowledge base, the global "repository of codes". Non-rivalry also makes knowledge cumulative, so that previous knowledge acts as basis for new ideas. As Isaac Newton famously put it in 1675: "If I have seen further it is by standing on the shoulders of Giants".

Therefore, the pace of technological progress is determined crucially by the size of the currently active research network. Even more precisely, as results obtained by researchers working in isolated communities don't diffuse into the wider world, the most adequate empirical measure of the scale operations in R&D is probably the size of the largest connected component of the global research network. The network perspective is particularly useful here because it helps distinguish between (i) growth in the number of active researchers and (ii) in the connectedness of global research networks. The former factor measures the raw research input, whereas the latter determines the efficiency in which the raw input is used for generating, incorporating, validating and disseminating new knowledge. Starting from a very low base, both elements grew over the millennia.

In the hunter-gatherer era, knowledge accumulation was pursued by small, disconnected bands and tribes, and hence there was massive duplication: the same knowledge (e.g., properties of plant and animal species, stone blades, making fire) was acquired independently by different hunter-gatherer tribes across the world, and this knowledge was hardly shared. Low connectedness of social networks in prehistorical times explains also why agricultural revolutions happened independently in various parts of the world, separated by millennia over which the useful information and technologies (e.g., domesticated species) did not diffuse beyond surprisingly limited geographical areas (Diamond, 1997). Even in the agricultural era, though, research networks of major populations were disconnected. For example, until 1200 CE, the exchange of technological ideas between China and Europe (and the Middle East) was virtually non-existent.

The scale of operations in R&D was dramatically increased at the time of the scientific revolution, accompanied by European geographical "discoveries" and colonial conquests which began in the late fifteenth century. Europeans' ruthless "guns, germs and steel" (Diamond, 1997) had also the side effect of gradually connecting the previously fragmented research networks, unifying the global pool of technological knowledge, and increasing the scale of operations in the now global R&D process. This reduced duplication externalities in knowledge creation (Mokyr, 2002). The concentration of power in the hands of Europeans also facilitated the emergence of a universal language of science. Moving away from Latin, it became a mixture of French, German and English, until it was fully dominated by English in the second half of the 20th century. At the dawn of the scientific revolution, the printing press empowered the humankind to share information much more broadly and establish a common methodological base for further discoveries and innovations. From then onwards, there was a systematic upward trend in the global number of active researchers as well as in the share of population and capital attributed to R&D actions (Mokyr, 2002, 2016).

The digital revolution, and in particular the Internet, provided another boost to the scale of operations in R&D by increasing the connectedness of global research networks. The recent decades have witnessed unprecedented increases in research cooperation and scientific output (e.g., research papers) per researcher. Researchers have begun to forge large-scale cooperations and consortia and build unique research facilities such as the Large Hadron Collider in Geneva or the Very Large Telescope in Chile. Our way to get around the cognitive limitations of the human mind, burdened by the exponentially growing knowledge base, is to narrow down our expertise and divide science into an increasing number of disciplines and research areas. Modern scientists are also helped in navigating the growing body of scholarly literature by the Internet with increasingly efficient search engines. Some of us may dislike the fact that the days of "Renaissance men" are long gone, and that specialists in different disciplines find it increasingly difficult to communicate with one another as well as explain their results to the broad public. However that's the only way the humankind as a whole can accommodate the fast-growing repository of available codes and the increasing complexity of these codes, and use it to extend our control over the world and better satisfy our needs (Hidalgo, 2015; Jones, 2009).

Given the digital-era emphasis on R&D, technological startups and innovation-led growth, I expect that the trends of increasing share of population and capital devoted to R&D, as well as increasing specialization, complexity and connectedness of global research, are likely to further intensify in the future (Bloom et al., 2020; Hidalgo, 2015; Jones, 1995).

6.4 Making of the Global Economy

Throughout the human history, increasing scale of operations in the creation of knowledge was accompanied by increasing scale of operations in the overall economy. The bands and tribes of the hunter-gatherer era, numbering from about 30–50 (typical size of roving bands) to about 150 individuals (the Dunbar's number), often connected by kinship ties, were generally disconnected and their "economies"—consisting primarily in the search and hunt for food—were largely autarkic. Interactions between neighboring populations were typically hostile. Barter trade between tribes was not unheard of, but was typically limited to scarce resources that were marginal for the survival and well-being of the tribe.

Agricultural production changed this picture. Transforming natural ecosystems to arable land and pastures generated food surpluses, which served to increase population size and fuel the rise of non-food-producing specialists. This resulted in gradual concentration of power over larger groups of people and geographical areas (Borcan et al., 2018; Lagerlöf, 2016). First centralized states emerged in the Fertile Crescent and surrounding areas where the agricultural technology was most advanced at the time. The earliest major city-states were founded in Mesopotamia—such as Uruk in the fourth millennium BCE (in today's Iraq) which may have had an urban population of 25 000 to 50 000 people—and the Indus Valley (in today's Pakistan). The first large-scale state was formed around 3100 BC in Egypt with the political unification of the entire lower Nile Valley, inhabited at the time by about 850 000–1 000 000 people. Ancient Egypt approximately tenfolded the area of the previous largest state.

Since that time, empires grew and fell, but the main underlying tendency was that later states were generally more populous than earlier ones, fueled by agricultural advances and gradual consolidation of power. The Achaemenid Empire, originating from the lands of modern-day Iran, came to rule over the entire Middle East, peaking at 49.4 million inhabitants around 480 BCE, or a staggering 44% of the world population. Around 1500 CE, the most populous state was China under the Ming dynasty, with about 125 million inhabitants, or 28.5% of the world population. Over time, economies of the agricultural era were gradually increasing their specialization in certain food crops and improving coordination of supplies within their territory. They also were slowly increasing the scale of their mutual interconnectedness by establishing long-range trade links. The most notable of those routes—the Silk Road—began around 130 BCE with the Han dynasty of China, connecting it to the Roman Empire as well as other adjacent regions in Central Asia and the Middle East.

The scientific revolution of sixteenth century Europe—accompanied with colonial conquests and later followed by the industrial revolution—was instrumental in establishing and cementing the European economic and military rule over most of the world. In particular, the British empire, whose growth was fueled by primacy in industrial development, eventually (in 1920 CE) came to rule over about 24% of Earth's land area and 23% of world population. Europeans were also the largest beneficiaries of booming international (and even intercontinental) trade. One of the most shocking examples of that phenomenon was the story of the British East India

Company, which was formed to trade in the Indian Ocean region, and later with China, but which—around 1800 CE—effectively seized control over large parts of the Indian subcontinent, colonized parts of Southeast Asia, and formed a private army and navy that played a major role in the victorious First Opium War against Qing China.

The history of European colonial conquests is a prime demonstration how technological progress and economic growth serve to increase the extent of control on behalf of their beneficiaries. It shows, in particular, that imbalances in access to technology can (and will) be easily transformed into imbalances in political power and economic affluence (Diamond, 1997). During most of the industrial era, a large part of the world was divided into European nation states and their overseas colonies.

Globally speaking, the industrial era was a time of rapid expansion in world population and output. According to Thomas Piketty's data (Piketty, 2014), in the period 1820–1990 CE world gross domestic product (GDP) grew at a rate of 2.14% per annum, doubling every 33 years. At the same time, global population grew by 0.96% per annum, doubling every 72 years. The industrial era witnessed a tremendous increase in the scale of economic operations. (Concurrently GDP per capita grew at a rate of 1.17% per annum, doubling every 60 years and contributing to a massive—though unequally distributed—rise in living standards.)

Growth in the connectedness of the global economy—that is, the extent of *globalization*—was equally staggering. Throughout the industrial era, more and more links were built between regions of the world, allowing not only for trade in commodities but also international migration and—increasingly—long-distance transmission of information, facilitated by the pathbreaking inventions of telegraph and telephone. Trade costs have decreased and volumes have soared, especially after a broad wave of trade liberalizations has swept across the world in the twentieth century, such as the General Agreement on Tariffs and Trade (GATT), first signed by 23 countries in 1947, later to be replaced by the World Trade Organization (WTO).

The digital revolution further accelerated the globalization process. Easier, cheaper and faster information transmission allowed international corporations to cut costs by offshoring parts of the production stream and establishing complex international networks of supply linkages. In the recent decades, we are observing a systematic trend of increasing fragmentation of production and creation of more and more sophisticated global value chains (GVC). It is not uncommon these days to buy a car of a Japanese brand, conceived and designed in Japan but assembled in Slovakia from components coming from Poland, Romania and France, with electronic parts shipped from Taiwan, and using raw materials like iron ore and oil mined originally in Australia and Russia. Modern information technologies are also instrumental in letting most efficient firms grow and become multinational corporations able to take advantage of increasing returns to scale at globalized markets.

In sum, over the course of the human history, the scale of economic operations has followed a clear upward trend from tiny autarkic economies, through warring city-states, competing but intensely trading nation states and colonial empires, to the modern strongly connected world economy that is permeated by increasingly intricate networks of business cooperation. The scale of an average economic unit rose from 30 to 50 members of a roving hunter-gatherer band, to the entire world population of about 8 billion people as of 2022.

This impressive development was achieved thanks to the human drive to maximize local control. This drive is a double-edged sword, though, capable of doing both great good and great harm. On the bad days, the struggle for control appeared as a zero-sum game (where one's good is other's harm), driving the humankind to world wars, ruthless colonial conquests, and genocide. On the good days, however, it promoted business cooperation, long-standing political alliances and establishment of mutually (and multilaterally) beneficial trade. All this helped forge economic linkages, join our efforts in creativity and innovation, tap the global potential of existing technologies, and exploit increasing returns to scale. Human local control maximization, as it turns out, is a positive-sum game after all: *on balance*, it enhances rather than stifles cooperation and makes our species better rather than worse off.

6.5 Institutions, Scale and Growth

Ever since Adam Smith's early investigations, the origins of "wealth of nations" have been a heavily disputed issue in economics. Why are some nations, countries, regions or groups of people better off than others? What determines the differences in their pace of growth? Despite centuries of interest, these questions are still not fully settled. Economists may have gained good understanding of proximate causes of economic growth, such as investment in physical and human capital, but the fundamental causes remain unclear. The key candidate explanations considered these days are geography, technology and institutions. In particular, biogeographical conditions, such as the availability of domesticable plant and animal species, have been shown to be important determinants of the timing of agricultural revolution across various parts of the globe (Diamond, 1997; Hibbs and Olsson, 2004). In turn, scientific and technological achievements as well as the favorable climate towards innovation during the European enlightenment have been shown to underly the fact that the industrial revolution was initiated in Europe rather than anywhere else (Galor, 2011; Mokyr, 2016).

But are *institutions*—such as rule of law, universal rights, control of corruption and violence, democracy—an important source of growth? Is the pace of national (and maybe even global) growth determined by state capacity and benevolence of the rulers? (Acemoglu et al., 2001; Borcan et al., 2018; Lagerlöf, 2016) It appears that it's not the case; instead it's primarily technological progress that underlies both economic and institutional development. The causality from institutions to technology and economic growth is observed only when the institutions are blatantly bad, contributing to prolonged economic stagnation or collapse.

A working model of technology and state institutions can be formulated with the following three building blocks. First, on principle, the level of technology determines the *maximum scale of state*, by which I mean a collection of maximally large forms of statehood that can survive for at least a few decades. In a world of small hunter-gatherer tribes, there was no possibility to establish and sustain statehood beyond the scale of the tribe. After the agricultural revolution, however, the scale of operations, both in economy and policy, systematically rose. First city-states were formed, followed by first centralized empires. The enlightenment and industrial eras further strengthened the centralized state in terms of tax collection and law enforcement compared to the agricultural-era powers of feudal rulers. At the time, it was technically feasible to sustain huge colonial empires and first representative democracies. Finally, after World War II, systematic attempts have been made to shift parts of authority from the national to the supranational and global levels.

The positive link between technology and the maximum scale of state follows from the fact that technological progress (i) eases information collection and transmission, and thus lowers surveillance costs, (ii) reduces costs and expands the capabilities of state bureaucracy and (iii) fuels economic growth and facilitates the pursuit of local control maximization. In result, on balance technological progress increases the incentives to stabilize authority, forge long-distance trade linkages and tighten cooperation in production and R&D, while lowering the incentives to fight and revolt.

Second, although technology defines the maximum scale of state, it doesn't determine its institutional setup. Technology doesn't decide whether the state is ruled as a totalitarian regime, a monarchy or a democracy, nor does it influence its economic and social policies. Such choices are a product of idiosyncratic historical, cultural, geographical and economic circumstances. For any technology level, however, there is a trade-off between *state strength* (military power, extractive capacity, centralization of power) and *inclusiveness* (empowerment of citizens to pursue their own local control). You may think of it as a choice problem where the product of state strength and inclusiveness is fixed by the level of technology, but the choice along the strength–inclusiveness axis is undetermined, so you may get both strong but extractive states, and smaller but more inclusive ones (Fig. 6.1). At the *maximum scale of state* frontier of 500 BCE, the centralized Achaemenid (Persian) Empire of about 5.5 million square km of land and 17–35 millions of inhabitants coexisted with the world's first democracy in the Greek city-state of Athens, populated by about 250–300 thousand people. Today the USA runs a representative democracy with about 325 million inhabitants, whereas China runs a centralized, authoritarian state of 1.394 billion—and nevertheless still manages to offer its citizens much less oppression and more safety and freedom than the (at least 40 times less populous) ancient Persian kingdom. Perhaps in the future, it will be possible to maintain stable global governance, running AI-augmented inclusive policies (Almossawi and Hidalgo, 2012) at the global scale—though the development of AI-augmented digital surveillance may also lead the world toward a digital dictatorship (Harari, 2017).

The trade-off between state strength and inclusiveness stems from the fact that strong states have a tendency towards oppression, which may give rise to internal resistance and amplification of centrifugal forces, whereas inclusive states have a tendency towards frailty and indecision, which makes them vulnerable to external conquests and internal *coup d'état*. One prediction of this theory is that in a cross-section of stable states, controlling for their wealth and access to technology,

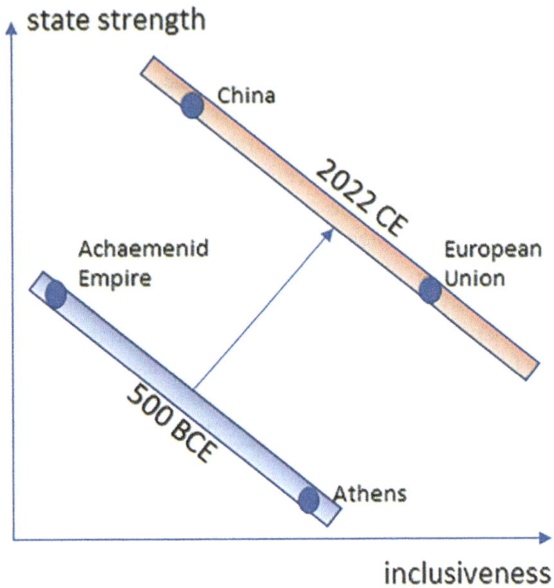

Fig. 6.1 Maximum scale of state and the state strength—inclusiveness nexus. *Note* This figure serves only for intuition building. The underlying theory has not been quantified yet. No clear numerical measures of state strength and inclusiveness exist

military power and centralization should be inversely correlated with inclusiveness and democratization. Over time, however, we should be observing a technologically driven trend towards both state strength and inclusiveness. Furthermore, extraordinarily large states—*too* large given the existing technology—should eventually become unstable and fall apart. Some argue that this is exactly what happened to the Roman Empire.

Third, multiple levels of governance may coexist. Large-scale states are often organized such that a lot of authority is passed in a top-down way to provincial leaders, regional and local governments and councils. For example, USA, India, Pakistan, Brazil, Russia and Germany are federal states. In the past, multi-level governance was also executed through a variety of feudal systems.

By contrast, the opposite direction of delegation—where smaller-scale entities voluntarily delegate parts of their authority to higher-level institutions—is rarely encountered, probably largely in fear of compromising the delegating authority's objective to maximize local control. It is not completely impossible, though. The European Union is a prime example of such voluntary delegation of power from nation states to supranational institutions such as the European Commission and the European Parliament. The EU was established—and later enlarged—to improve cooperation, exploit increasing returns to scale at the European level, broadly share the ensuing gains, and firmly establish incentives preventing the outbreak of another war in Europe. It is however an open question if the EU, strongly tilted towards inclusiveness on the strength–inclusiveness spectrum, will prove sufficiently strong to survive against mounting centrifugal forces such as the Brexit. Apart from the EU, however, thus far we haven't observed authorities giving up power to larger entities other than being militarily conquered or otherwise economically or politically forced.

This working model justifies my belief that over the long run, efficient state institutions are not a key prerequisite for technological progress and economic growth. Instead, state institutions are selected from the set of technological possibilities along the state strength–inclusiveness spectrum. Nevertheless, adverse (sub-par in relation to the "frontier") political circumstances may have hindered technological and economic growth in certain periods such as the early Middle Ages. Furthermore, in a cross-section of countries, failed states (with near-zero state strength) and unjustifiably tyrannical regimes (that offer too low inclusiveness given their strength) are notoriously correlated with low incomes and technological backwardness.

Finally, it has been speculated in the historical literature that inappropriate political institutions may have resulted in certain development possibilities being missed. This appears to be a very plausible hypothesis. For example, it remains a puzzle why China under the Ming dynasty did not initiate the scientific revolution in fourteenth to fifteenth century CE, before Europe did. After all, already during the Song dynasty (tenth to thirteenth century CE)—i.e., in parallel with European Middle Ages—China was the world's unquestionable technology leader. It also had a markedly greater population—125 million inhabitants in 1500, compared to 90 million in the whole of Europe combined—and was better equipped to exploit increasing returns to scale thanks to being politically unified. China seems to have had everything in its own hands, so why did it miss this opportunity? We don't know exactly, but according to Mokyr (2016) the key reason may have been institutional: "[i]n Europe, heterodox and creative thinkers could find sanctuary in other countries and spread their thinking across borders. In contrast, China's version of the Enlightenment remained controlled by the ruling elite". Apparently, Yuval Noah Harari's observation that "politicians have a very limited capacity to do good, but an unlimited capacity to do harm" may apply also to doing harm through inaction.

6.6 A Case for Global Digital Policy

Now, fast forward to the present time and the near future. It is an interesting thought experiment to juxtapose the above working model of technology and state institutions with the recent developments of the digital era. There is a case to make that modern digital technologies have markedly increased the maximum scale of state (pushed the state strength–inclusiveness "frontier" upwards). First, they ease information collection and transmission, and thus lower surveillance costs. Second, they tend to reduce costs and expand the capabilities of state bureaucracy. Third, they also reinforce integration of the global economy and tighten up interdependence through global value chains.

Thus far, this increase in the maximum scale of state generally hasn't been reflected in changes of policy. (Save, perhaps, for China growing both stronger and more inclusive over the last decades. China is also known for being the first to experiment with gamification of social status and to deploy cutting-edge AI for surveillance.) Almost all of our running economic policies and institutions have been designed with the industrial-era economy in mind and not redesigned since. Therefore, as the digital era progresses, they are becoming increasingly inefficient. And dedicated digital-era economic policies or institutions have not been devised yet: compared to the industrial economy, the digital economy is *virtually unregulated*.

Why is it so? Well, it shouldn't come as a surprise when we look at the timing. As of 2022, we are arguably still at an early stage of the digital era; after previous technological revolutions it also took time to develop suitable policies and institutions. When the agricultural revolution of the Neolithic period allowed for accumulation of food surpluses, this had initially rather anarchic outcomes until the state emerged and enforced hierarchical order, much more centralized than in the previous hunter-gatherer era (Diamond, 1997). Analogously, the industrial revolution unleashed fast-growing, *laissez-faire* industrial capitalism which was only gradually regulated later by installing worker rights, corporate taxes, public education, health insurance, social security, etc. Ultimately, in the Western world, this led to the welfare state with an increasingly high share of state budget in GDP.

There are many signals that the digital-era economy is very loosely regulated compared to the industrial-era economy. Authorities are not, e.g., taxing incremental data flows (which would be a digital-era equivalent of VAT), taxing software providers for their returns to data collection (which would be an equivalent of corporate income tax), enforcing legal penalties for distributing harmful data, etc. On the one hand, this offers great freedom and perspectives of untamed development of the global digital economy. On the other hand, negative consequences are also abound. For example, there exists a sizable "dark web", a digital-era representation of the shadow economy. Cyberpolicing is lagging behind cybercrime. Organized Internet-based misinformation campaigns are able to affect democratic voting outcomes and public sentiments but so far cannot be traced back to their original organizers. There is an emerging arms race towards most versatile and adaptive AI technologies with relatively little research on AI safety. Insufficient regulation arguably also facilitates the formation of global monopolies, large-scale intrusion into privacy (according to Zuboff (2019), paramount to establishment of *surveillance capitalism*), and generally the potential to abuse the position of power on the side of large technological corporations. Giant software corporations like Google, Facebook and Amazon in the US and Baidu, Tencent and Alibaba in China, are probably able to execute their local control to an extent way exceeding their market valuation. They are able to do so because as long as we can't really estimate the value of information and fully appreciate the power of increasing returns to scale in an automated economy, they can appropriate the associated positive externalities to themselves.

The fundamental reason why the digital-era economy is virtually unregulated is that contemporary policies—and arguably also state institutions—are not adapted to the unprecedented scalability of the global digital economy. The rapid buildup of programmable hardware in the last decades, progress in development of pre-programmed software and AI, and acceleration of data collection have allowed technological corporations to exploit increasing returns to scale much better than ever before. Software applications can be immediately deployed to the entire global market. Start-up companies can become major global players very quickly, and thus—knowing that—incumbent companies have strong incentives to act preemptively and fill the emerging market niches first or buy out the newcomer. For example, Alphabet Inc., a holding company set up by Google has already acquired more than 200 companies, including Android, Waze and DeepMind.

Furthermore, the digital-era economy is decidedly global. Costs of "relocating" digital business are historically small: it is much easier to re-route data traffic or change the legal domicile of intangible capital, than to re-route material goods or establish overseas branches. Digital-era corporations rationally exploit this fact, not just to save on taxes but also to avoid costly legal restrictions on data collection and communication in some countries (for good or bad—these include both state censorship and protection of intellectual property and privacy). And state bureaucracies just can't keep up with the growing complexity of a globally interconnected world where the volume of accumulated data doubles every 3 years.

In the eternal race between private enterprises and policy-makers, firms always have the first-mover advantage and put effort to exploit newly arising opportunities before these are curtailed by public policy. What is new in the digital era, though, is that the accelerated pace of growth increased the social cost of policies being introduced with a lag.

Designing better policies and institutions for the digital era requires to ask some important questions. First, are we able to reliably trace and quantify the objects of interest, such as social returns to software, or the value of flows of new non-redundant data? (Note that in the early years of the industrial revolution, when only agricultural-era regulations and policies were in place, physical capital and industrial output were also viewed as rather fluid, elusive concepts, not so easy to quantify and regulate as land ownership and crop yields; now we do this on a daily basis.) If not, we should quickly learn how to do it (see Mazzucato (2018)).

Second, can we devise sensible digital policy that would adequately address the emerging challenges? One of the challenges is that given the global character of the digital economy, such policy should arguably be also globally administered. (Global cooperation is paramount here because what is being accomplished for the industrial economy by tax havens, will be only much more cheaply and easily—accomplished for the digital economy by "data havens".) Who should design and run this new global administration and would it be publicly legitimized? Some global organizations like the World Economic Forum have come ahead and recognized many promises and perils of the digital revolution (Schwab, 2016), but the WEF is an international organization for public-private cooperation, not a legal authority. Should such authority be established, for example, under the auspices of United Nations?

Third, do we have enough time to wait patiently until the world accepts that there should be legitimate authorities at the global level, even if specializing only in digital policy? There are reasons to be pessimistic here, because (i) there is very scarce historical evidence of national authorities giving up part of their power to supranational entities—essentially only the European Union; (ii) otherwise, authority was transferred upwards only due to military or economic conquest; (iii) the technology and economy are changing at an unprecedented pace, making it hard for people to mentally keep up.

The bottom line is that the institutional implications of digital-era technologies have surpassed the boundaries of our imagination. Most of us still think at the local and national scale, and our policies and institutions remain aligned with our obsolete mindsets. Our perception of the world lags behind reality by about one generation, or 30 years (Rosling, 2018). Similarly, the mainstream of economics as a science, and most economic thinking in general, is still rooted in the industrial era. Some say, for example, that modern digital technologies constitute just one more wave of the industrial revolution ("Industry 4.0") whose effects are already fading out (Gordon, 2016). Only that it's not true. The sooner we understand and internalize the new reality, the more time we'll have to devise appropriate policies.

References

Acemoglu, D., & Robinson, J. (2012). *Why nations fail: Origins of power*. Poverty and Prosperity: Crown Publishers (Random House).

Acemoglu, D., Johnson, S., & Robinson, J. (2001). The colonial origins of comparative development: An empirical investigation. *American Economic Review,91*, 1369–1401.

Almossawi, A., & Hidalgo, C. (2012). Going beyond the one-bit democracy. *Wired,11*.

Bloom, N., Jones, C. I., Van Reenen, J., & Webb, M. (2020). Are ideas getting harder to find? *American Economic Review,110*(4), 1104–1144.

Borcan, O., Olsson, O., & Putterman, L. (2018). State history and economic development: Evidence from six millennia. *Journal of Economic Growth,23*, 1–40.

Comin, D., & Hobijn, B. (2010). An exploration of technology diffusion. *American Economic Review,100*, 2031–2059.

Diamond, J. (1997). *Guns, germs, and steel: The fates of human societies*. Norton & Company.

Dunbar, R. I. M. (1998). *Grooming, gossip, and the evolution of language*. Harvard University Press.

Galor, O. (2011). *Unified growth theory*. Princeton University Press.

Gillings, M. R., Hilbert, M., & Kemp, D. J. (2016). Information in the biosphere: biological and digital worlds. *Trends in Ecology and Evolution,31*, 180–189.

Gordon, R. J. (2016). *The rise and fall of American growth: The U.S. standard of living since the civil war*. Princeton University Press.

Harari, Y. N. (2014). *Sapiens: A brief history of humankind*. Vintage.

Harari, Y. N. (2017). *Homo deus: A brief history of tomorrow*. Vintage.

Hibbs, D. A., & Olsson, O. (2004). Geography, biogeography, and why some countries are rich and others are poor. *Proceedings of the National Academy of Sciences,101*, 3715–3720.

Hidalgo, C. (2015). *Why information grows: The evolution of order from atoms to economies*. Basic Books.

Jones, C. I. (1995). R&D-based models of economic growth. *Journal of Political Economy,103*, 759–84.

Jones, C. I. (2005a). Growth and ideas. In P. Aghion & S. Durlauf (Ed.), *Handbook of economic growth*.

Jones, B. F. (2009). The burden of knowledge and the "Death of the Renaissance Man": Is innovation getting harder? *Review of Economic Studies,76*, 283–317.

Kremer, M. (1993). Population growth and technological change: One million B.C. to 1990. *Quarterly Journal of Economics,108*, 681–716.

Lagerlöf, N.-P. (2016). Statehood, democracy and preindustrial development. *Journal of Economic Dynamics and Control,67*, 58–72.

Mazzucato, M. (2018). *The value of everything: Making and taking in the global economy*. Public Affairs Books.

Mokyr, J. (2002). *The gifts of Athena: Historical origins of the knowledge economy*. Princeton University Press.

Mokyr, J. (2016). *A culture of growth: The origins of the modern economy*. Princeton University Press.

Pacho, W. (2013). *Stagnacja gospodarcza i innowacje techniczne w okresie przedindustrialnym*. Oficyna Wydawnicza SGH.

Piketty, T. (2014). *Capital in the twenty-first century*. Translated from French by Arthur Goldhammer. Harvard University Press.

Romer, P. M. (1990). Endogenous technological change. *Journal of Political Economy,98*, S71–S102.

Rosling, H. (2018). *Factfulness: Ten reasons we're wrong about the world—and why things are better than you think*. Sceptre.

Schwab, K. (2016). *The fourth industrial revolution*. World Economic Forum.

West, G. (2017). *Scale: The universal laws of growth, innovation, sustainability, and the pace of life in organisms, cities, economies, and companies*. Penguin Press.

Zuboff, S. (2019). *The age of surveillance capitalism: The fight for a human future at the new frontier of power*. Public Affairs.

Part III
The Great Leaps

Chapter 7
Accelerating Growth in the Past

Inventions have long since reached their limit, and I see no hope for further developments.

Roman engineer Julius Sextus Frontinus, 10 CE

Before man reaches the moon, your mail will be delivered within hours from New York to Australia by guided missiles. We stand on the threshold of rocket mail.

Arthur Summerfield, U.S. Postmaster General, 1959

7.1 What Is Development?

Over the course of the human history, actions of our ancestors have brought us from being just one of many animals in their struggle for survival to a world organized so that the *homo sapiens* can satisfy its needs extremely easily. No other species is able to stand in our way and contradict our control anymore. The five major technological revolutions were the key stepping stones of this transition, great leaps in global "economic development" serving the *homo sapiens*. But what *is* development after all? How to measure it consistently at such a long time scale?

If one accepts the proposition that development of the human civilization is a consequence of the human drive to maximize local control, a direct way to measure development would be to calculate the grand *sum of human control*. But then again, what could this mean in practice? We need something that is easier to handle empirically. And here's where problems begin. All proxy variables one can think of have problems of one kind or another, whether they are, e.g., global population size, global GDP, aggregate optimization power of human brains and man-made machines, the

J. Growiec, *Accelerating Economic Growth*, Frontiers in Economic History,
https://doi.org/10.1007/978-3-031-07195-9_7

aggregate stock of stored data, multidimensional measures of the degree of satisfaction of certain human needs, self-reported well-being, life satisfaction or happiness.

Economists' favorite methodology for measuring economic development at the regional, sectoral, national and global level is the System of National Accounts (SNA). The SNA has been constructed—first in 1953 and then subsequently revised—to ensure coherent measurement of all economic activity, maintaining comparability across sectors, countries and time: not only to meaningfully compare apples and oranges but, say, apples in Poland in 1997 and oranges in Greece in 2015. What comes out of national accounts as a summary measure of global development is the annual gross domestic product (GDP), or gross value added (GVA), expressed in common currency units (e.g., dollars, euros, or Purchasing Power Parity units, PPPs) and summed over countries.

I'm not going to criticize this approach here, despite some legitimate concerns (Mazzucato, 2018; Stiglitz et al., 2009). I think that national accounts are an excellent way of measuring *economic development* in the narrow sense of the term. In particular, there probably isn't anything better to measure the level and pace of economic development over the last decades. Problems begin to mount, however, when handling incomplete data from the further past or trying to extrapolate the data into the future. Dealing with the times before national accounts and national statistical offices existed, economists are forced to construct proxy variables based on the available information on population size, tax revenues, prices of traded and non-traded goods, etc., and use some creative interpolations. This is precisely what is being accomplished, among others, by Angus Maddison and his excellent successors of the Maddison Project at the Groningen Growth and Data Center, as well as Stephen Broadberry and his collaborators at Oxford. In turn, looking into the future one may wonder if national accounts don't underestimate the change in scope of human local control and satisfaction of our needs due to information technologies. For example, time use studies find that digital goods offered over the Internet may affect our lives and the global economy to a greater extent than their dollar value would suggest (e.g., Brynjolfsson and Oh (2012)).

All in all, taking a step back and looking at the whole human history suggests that the national accounts provide a rather incomplete characterization of development of the human civilization at the global scale and in the very long run. They provide *exact* measurement but only of *part* of our actual development. There is room for new insights from a more comprehensive approach.

7.2 New Dimensions of Development

To get a broader view of development processes, I propose a two-level approach: firstly, to keep aggregate human control as a unique overarching frame, but then, secondly, to break it down into era-specific measures of development. This approach formalizes the idea that *each new era opens an entirely new dimension of development*, that operates on top of the earlier ones and exhibits much faster growth.

Table 7.1 Decomposition of the level of development, including only hardware

Q	$=$	Hab	$\times$	$\frac{POP}{Hab}$	$\times$	$\frac{GDP}{POP}$	$\times$	$\frac{Q}{GDP}$
		Hunter-gatherer		Agricultural		Industrial		Post-digital

Growth in the new dimension not only adds to the growth in the earlier dimensions, but also—thanks to certain positive feedback effects that I'll discuss later—strongly accelerates it. (To my knowledge, this is a new perspective.)

I propose two alternative breakdowns of aggregate development, understood as the sum of human control: a fourfold decomposition focused only on hardware, and a sevenfold decomposition including both hardware and software.

In terms of hardware alone, aggregate development (denoted as Q) can be split into components related to: habitat capacity (Hab), population (POP), GDP, and a residual, interpreted as aggregate development per unit of GDP. The logic of this breakdown is as follows. First, in a hunter-gatherer economy (both before and after the cognitive revolution) the extent of human control could be identified with the total carrying capacity of ecosystems under human rule, Hab. Second, in an agricultural economy, characterized by Malthusian population dynamics, the extent of human control was in turn equivalent to total human population, POP. Third, after the industrial revolution the extent of human control can be identified with world GDP. Finally, we shouldn't rule out the possibility that new hardware revolutions will appear in the future, providing new means of exercising local control. All these hypothetical future developments are lumped in the residual component. The resulting equation is as presented in Table 7.1:

Table 7.1 should be read from left to right and interpreted in the following way. First, fix a moment in time, say 1700 CE, about a century before the industrial revolution. Then all the "yet inactive" hardware dimensions that will be opened only after this date (in this case, GDP/POP and Q/GDP) are fixed at unity. In consequence, the aggregate level of development is identified with the last open dimension (here, $Q \approx POP$) and decomposed into a relevant number of components (two), describing the period from the inception of the *homo sapiens* to the given date. For 1700 CE, the two active components are: habitat capacity (the hunter-gatherer measure of development) and population sustained relative to the natural habitat capacity (the intensive margin of development opened by agriculture).

The above decomposition is useful also for understanding the proximate sources of growth. Take a ratio of Q at two distinct points in time, say 1700 CE and 2000 CE. The equation then breaks down growth in Q into four components: (1) growth in conquered habitat capacity—of relatively minor importance but there have been some gains from, e.g., colonization and introduction of European crop species and farming techniques in previously sparsely inhabited regions of the Americas; (2) growth in population size relative to habitat capacity—of major importance as the considered period was a time of population explosion, fueled by progress in medicine, mechanization and modernization of agriculture, introduction of new crops, etc.; (3) growth in GDP per capita—of enormous importance as the industrial revolution appeared between both dates, unleashing rapid growth in labor productivity and standards of living; and (4) growth in Q per unit of GDP—a yet inactive dimension.

An implicit assumption in this reasoning is that a new technological revolution can occur only if the previous-era development level is sufficiently high. The new era always operates *on top of* the previous one: there would have been no agricultural revolution if the natural habitats had not been conquered by the earlier hunter-gatherers and no industrial revolution without secure supply of food for a large (and growing) population. Analogously, we shouldn't expect any future "post-digital" hardware revolution without reliable supply of both food and a wide variety of industrial and digital-era goods and services. The consecutive revolutions can also be viewed through the lens of Maslow's (1954) hierarchy of needs: each new era opens new possibilities for satisfying higher-level needs, but pursuing them is possible only if lower-level needs are already sufficiently satisfied.

But what of the software revolutions: the cognitive, scientific and digital revolution? Haven't they also affected the scope of our control? Shouldn't we include them in the decomposition as well? Well, to me the best answer to this question is "it depends". It depends, namely, on whether we're interested only in the tangible effects of our control or are ready to include the informational aspects, too. Intuitively, aggregate development Q must be at least partly tangible, serving our bodily needs at the bottom of Maslow's pyramid. But it may just as well include also intangible outputs serving our higher-level psychological needs.

An alternative breakdown of development Q includes components related to: habitat capacity (Hab), ecosystem information (EI), population (POP), scientific knowledge (SK), GDP, useful data (UD), and ultimately a residual, interpreted as aggregate development per unit of useful data. The logic of this decomposition is as follows. First, before the cognitive revolution the extent of human control was equal to the total carrying capacity of ecosystems under human rule, Hab. Second, after the cognitive revolution, when the humankind passed the threshold of net positive collective knowledge accumulation, we slowly began to build up our knowledge of the surrounding ecosystem, EI—the plants, the animals, how sticks and stones could be used to form tools, etc. This information was paramount in allowing our ancestors to domesticate certain species and initiate agriculture. Third, in the agricultural era, the extent of human control was equivalent to total human population, POP. Fourth, after the scientific revolution of the sixteenth century, the humankind began collecting scientific knowledge, SK, and systematically distinguishing it from false theories, undocumented beliefs and myths. This information was paramount in setting the stage for the following industrial revolution. The fifth component is GDP, the best available measure of development in the industrial era. Sixth, after the digital revolution of 1980 and beyond, we observe a surge of digitally stored information, useful data UD, that is also, at least partially, instrumental in furthering our control and serving our needs. The seventh and final component is the residual, covering all hypothetical post-digital developments. Hence the full decomposition, marking the five historical technological revolutions and one hypothetical one, is given in Table 7.2:

Decomposition Table 7.2 should be interpreted analogously to the former "hardware only" decomposition. First, fix a moment in time (say, 1700 CE). Then all the

Table 7.2 Decomposition of the level of development, including both hardware and software

$Q=$	Hab	$\times\frac{EI}{Hab}$	$\times\frac{POP}{EI}$	$\times\frac{SK}{POP}$	$\times\frac{GDP}{SK}$	$\times\frac{UD}{GDP}$	$\times\frac{Q}{UD}$
Era	Pre	Cogn	Agr	Sci	Ind	Dig	Post
	X	S	X	S	X	S	X
Time	−200 k	−70 k	−8 k	1550	1800	1980	?

Notes Pre—the period prior to the cognitive revolution, when *homo sapiens* occupied the middle of the food chain; Cogn—hunter-gatherer era after the cognitive revolution; Agr—agricultural era; Sci—enlightenment era following the scientific revolution; Ind—industrial era; Dig—digital era; Post—hypothetical post-digital era; *X*—hardware, *S*—software

dimensions that will be opened after this date (in this case, GDP/SK, UD/GDP and Q/UD) are fixed at unity. In consequence, the aggregate level of development is identified with the last open dimension (here, $Q \approx SK$), and decomposed into a relevant number of components, equal to the number of past technological eras since the inception of the *homo sapiens*. In our case, there are four such components: Hab (habitat capacity), EI/Hab (the intensive margin describing how well humans know their ecosystem), POP/EI (how well humans are able to apply their knowledge of nature to feed themselves and increase the population), and SK/POP (the average volume of scientific knowledge per person).

A problem with the above decomposition, however, is that it is unclear if one could really disentangle ecosystem information from scientific knowledge and (digital) useful data. After all, this is all information of one sort of another, isn't it? As new discoveries and innovations come along, we put the new facts beside the old ones. The only difference between EI, SK and UD is in the way the information is collected, processed and stored. But is it an important difference? Some pre-historical ecosystem information—for example that champignon mushrooms are edible whereas flybanes are deadly—is just as valid today as it was millennia ago. Should the ancient knowledge of celestial trajectories of the Sun, Moon and stars be classified as "ecosystem information", but after Copernicus built a theory that explained these paths—reclassified as "scientific knowledge"? And what about the time series of prices of stocks listed at the New York Stock Exchange: are they "scientific knowledge" or "useful data"? Assuming unity of knowledge (Harris, 2010), the lines between EI, SK and UD appear arbitrary.

If you think that this problem is important (and I am still hesitating), then you may find it more valuable to use the "hardware only" decomposition of aggregate development and only keep in mind that the consecutive hardware revolutions were made possible by the underlying trend of accelerating growth in acquisition, communication and processing of information. Nevertheless, singling out the UD/GDP component alone teaches us a valuable lesson about the contemporary digital era. The digital sphere is now growing an order of magnitude faster than GDP, and although the effects of this growth have an intangible, informational character, they are also providing demonstrable benefits for our well-being and the extent of our local control. We'd better not throw the baby out with the bathwater by disregarding the digital revolution in our calculations of the global level of development.

7.3 Revolutions in Energy, Revolutions in Learning

In the Introduction, I have argued for a developmental pendulum that alternatingly produces breakthroughs in the realms of hardware and software (Fig. 1.2). There is, however, an alternative way to look at these breakthroughs: respectively as revolutions in *energy* and in *learning*. After all, what made these great leaps in human history so unique was not that they provided us with additional physical action or with additional useful information. No, the key was that they empowered us with new sources of energy for performing the actions and, respectively, new ways of learning how to accumulate and process information. They gave us the fishing rod, not just the fish.

Hardware revolutions are crucially "energy revolutions", great leaps in the humankind's ability to harness energy and use it to perform actions that manifest our control. The agricultural revolution initiated around 10 000 years ago empowered our ancestors to override natural growth processes, transform ecosystems into farmlands that produce much more calories of energy per land area, and in consequence feed a much larger human population. The industrial revolution of ca. 1800 CE, in turn, empowered humans to release the energy captured in fossil fuels. The most fundamental innovations of the industrial revolution were the steam engine, the internal combustion engine and electricity. The only difference between these two revolutions is that while our human bodies run on food, our machines usually run on oil, gas, coal and electricity, itself produced largely from oil, gas, coal and uranium.

Software revolutions, by contrast, are "learning revolutions", great leaps in the humankind's ability to accumulate, communicate and manipulate information. They don't directly increase the level of development in the tangible, material sense, but—since information is non-rivalrous and therefore a source of increasing returns to scale (Romer, 1990)—indirectly prepare the ground for next hardware revolutions. Approximately 70 000 years ago, the cognitive revolution initiated the process of collective learning among the human species. For the first time in Earth's history, information could be passed and systematically accumulated across generations without the need to be imprinted in the species' genetic code. The scientific revolution of ca. 1550 CE rapidly accelerated human learning and the accumulation of knowledge thanks to the printing press (which greatly facilitated dissemination) and the scientific method (which greatly improved our ability to distill facts and working theories from false beliefs and myths). The recent digital revolution, in turn, was the first one in the sequence of software revolutions that detached information storage, processing and communication from the human brain. As scientists and engineers developed digital memory, data processing units, and eventually the Internet, the process of learning transgressed the narrow confines of our heads, allowing the human civilization to enter an era of *machine learning*.

This perspective suggests that may be one could limit the list of global development indicators to just two: the volume of energy used by our species and the volume of collected data. Making such an approach useful would, however, require a lot of work—among other issues—to filter out energy that is wasted and information that is redundant.

Table 7.3 New resources accessed thanks to technological revolutions

Revolution		Key breakthrough	New dim.	Broad access to
Cognitive	S	Increased human brain capacity	EI	Habitats
Agricultural	X	Plant domestication	$\frac{POP}{Hab}$	Food (calories)
Scientific	S	Books and scientific method	SK	Knowledge
Industrial	X	Harnessing energy from fossils	$\frac{GDP}{POP}$	Physical goods
Digital	S	Programmable computers	UD	Information

Note X - hardware, S - software.

There is one more perspective from which one could also look at the technological revolutions: what have they broadened our access to? What resources had been previously inaccessible or available only to a select few, but later spread rather widely? This is a perspective that addresses not just the increase in the total volume of produced output—however measured—but also its distribution among the human population (Table 7.3). (Please note that information measures are not divided by the previous-era development measure because information is non-rivalrous.)

Thanks to the slight but pivotal increase in human brain capacity acquired during the period of cognitive revolution, our hunter-gatherer ancestors passed the threshold of net positive cumulative accumulation of knowledge. They could finally effectively *learn*—and the key thing to learn was how to acquire control over new habitats which were previously occupied by other species. They used this knowledge to eliminate their enemies, advance up the food chain (Harari, 2014) and spread around the world. Shared ecosystem information was a powerful mechanism democratizing human control over natural habitats.

All the subsequent progress only increased the pace of accumulating information about the properties of the surrounding plants and animals. In particular, our ancestors learned that certain cereal species such as wheat or barley have a high caloric content and can be grown by planting seeds. Putting such ideas in practice led to the agricultural revolution (or more precisely, at least seven independent agricultural revolutions across the globe, separated by millennia). Domestication of food crops and transforming natural habitats into farmlands opened up the intensive margin of increasingly efficient production of *energy* from a unit of land, allowing to sustain more human lives per square mile. Successful development could then be done not just by conquering new land but also by turning already acquired land into farmlands and intensifying agriculture (Diamond, 1997).

The European Renaissance demonstrated the implications of the fact that economic prosperity is rivalrous but knowledge isn't. The fifteenth century invention of the printing press, coupled with the increasingly common ability to read, massively broadened access to knowledge. Affordable printing of books at scale shielded human knowledge from being forgotten and allowed scholars to gradually specialize.

A century later Europe witnessed broad adoption of the scientific method and rapid buildup of scientific knowledge, including astronomy, physics, chemistry, medicine, geography, biology, and even social sciences like economics. The humankind really learned a lot.

One important thing our ancestors learned at the time was how to harness energy by burning fossil fuels. This paved the way for the industrial revolution. From the steam engine to electricity, the technological breakthroughs of that era greatly improved the humankind's potential for performing controlled physical actions. Soon there was a multitude of ideas around what kinds of actions would be most profitable. For example, in the early nineteenth century Manchester, England became a global center of textile production, fueling its newly developed steam-powered weaving machines with local coal, and transporting its products to the sea port of Liverpool on the newly opened steam-powered railway line. The industrial revolution initiated the accumulation of physical and human capital (educational attainment and skills), and opened up the intensive margin of growth in consumption per capita—that is, allowed more and more goods (in terms of quantity, quality and variety) to be consumed per person. After the initial period of rapid buildup of physical capital, industrial-era economies began to exhibit an ever-growing demand for human cognitive work. Due to the underlying children quantity–quality trade-off this triggered a decline in fertility, thus finally breaking the Malthusian mechanism that had been affecting our species to that day (Galor, 2011; Galor and Moav, 2002). In the end, in the industrial era aggregate human control could be increased either extensively (via population growth) or intensively (via growth in GDP per capita). Thanks to growing wages, since the second half of the nineteenth century industrialized countries in Europe and America witnessed a clear pattern of democratization of access to physical goods and political representation, and a gradual decline in income inequality. In the twentieth century, industrialization and its achievements began diffusing around the world.

In the meantime, the industrialized world was increasingly investing in R&D. In particular, R&D was increasingly carried out by private enterprises in addition to public universities and research institutes. And even if some of the research was aimed at driving competitors out of the market or winning a war against a different country (both of which are zero-sum or even negative-sum games), on balance the R&D boom was extremely beneficial for the human civilization. By increasing the variety and quality of consumed goods and reducing costs of production, the aggregate extent of human control increased substantially, and so did our ability to satisfy our needs.

But the greatest results of learning are achieved when the process of learning is itself improved. That's exactly what happened in the second half of the twentieth century. The digital computer constructed by Alan Turing and his collaborators in the 1940s immediately after World War II was the first piece of hardware that could perform "cognitive" (in this case, arithmetic) operations as well as store and run its own code. (Alan Turing was to the digital revolution what Johannes Gutenberg was to the scientific revolution: inventor of the critical enabler of the later breakthrough.) After a series of cumulative improvements of Turing's brilliant invention, capabilities of digital devices skyrocketed. Since the 1980s progress in data processing power, storage capacity, miniaturization of silicon chips, and communication bandwidth came to exhibit Moore's law—unprecedented pace of exponential growth, with doubling periods of approximately 1.5–3 years.

Since the dawn of the digital era, printed books and journals are becoming obsolete, and so is—to a large extent—our own memory. Learning is getting more and more detached from the human brain. Nevertheless, at the same time access to information, or useful data (UD), is becoming increasingly broad-based and democratic. It has never been so easy to obtain and spread data—whether true or false, signal or noise—and people are systematically taking advantage of this, for example by spending many hours a day browsing the Internet or creating online content based on our own thoughts and ideas. It appears that we tend to draw more utility from useful data than its pecuniary value would suggest (Brynjolfsson and Oh, 2012; Rosling, 2018). Hence, one may say that in the digital era aggregate human control can be increased either extensively (via GDP growth) or intensively (via growth in useful data per unit of GDP).

An important implication of the fact that the digital revolution has decoupled information processing from human brains is that in the future, the share of consumption of material output that serves to sustain programmable hardware (computers, robots, smartphones, etc.) rather than the human population may one day shoot through the roof. Already today an increasing share of productive software is embedded in non-human hardware rather than human bodies, and there is a possibility that after more and more processes become fully automated, our cognitive skills may one day become completely irrelevant for production.

7.4 Accelerations? By Orders of Magnitude!

So how strong were the growth accelerations following the five past technological revolutions? Can we put any numbers into Tables 7.1 and 7.2?

The answer is yes. Thanks to the tedious work of economic historians such as Angus Maddison who scanned through a plethora of original sources and invented many ingenuous approximations, we now have reasonably good knowledge of past economic development at the global level. The data confirm that each new era opened a new dimension of development that started to expand at an exponential pace that was *at least an order of magnitude* faster than growth in the previous dimensions (Table 7.4, Fig. 7.1). Over the course of human history, growth was massively super-exponential: the doubling period for aggregate human control Q has shrunk from 34 360 years in the hunter-gatherer era to 885 years in the agricultural era (as measured by population growth), 33 years in the industrial era (global GDP growth), and just 2.8 years in the digital era (growth in the volume of communicated data).

What stood behind these impressive accelerations was the great pendulum of technological and economic development, restlessly swinging between hardware and software. Hardware revolutions allowed the humankind to harness more and more energy, which led to new forms of physical actions which improved economic prosperity. Software revolutions, in turn, increased the pace of learning, which boosted

Table 7.4 Estimates of the pace of economic growth across the eras

Era	Period	Measure	Growth (%)	Doubling
Hunter-gatherer	25 000–5 000 BCE	Population	0.002	34 360 y
Agricultural	5 000 BCE–1500 CE	Population	0.069	1007 y
Enlightenment	1500 CE–1820 CE	Population	0.271	256 y
		... world GDP	*0.324*	*214 y*
Industrial	1820–1990 CE	*... population*	*0.962*	*72 y*
		World GDP	2.14	33 y
Digital	1990–2012 CE	*... population*	*1.30*	*54 y*
		... world GDP	*3.41*	*21 y*
		Data communication	28	2.8 y

Sources Kremer (1993a), Piketty (2014), Hilbert and López (2011). See the details in text.
Note time brackets for the respective eras are approximate due to problems with data availability. There is no reliable data on total habitat capacity Hab or total population of *homo sapiens* prior to the cognitive revolution

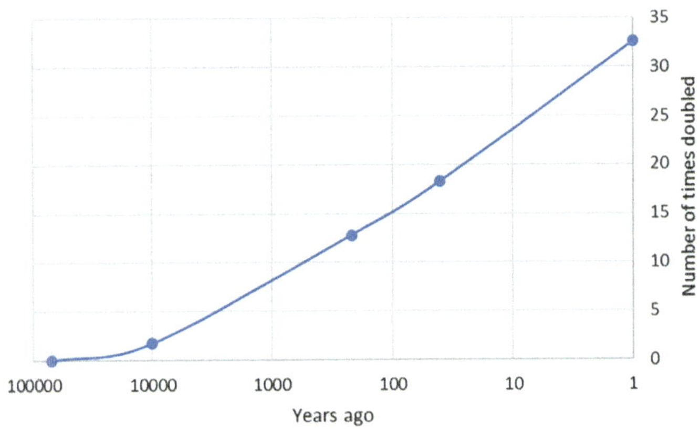

Fig. 7.1 The cumulative number of times the level of development has doubled. *Sources* Kremer (1993a) for data points before 0 CE, Piketty (2014) otherwise. *Notes* dots denote the approximate timing of technological revolutions (agricultural, industrial, digital). Time is presented in logarithmic scale to mark that technological revolutions have been becoming closer to one another in time

returns to research effort, accelerated the diffusion of information and increased the scale of operations in R&D, ultimately leading to new breakthrough ideas that underlay the consecutive revolutions in hardware.

Economic growth in the hunter-gatherer era after the cognitive revolution was driven by two factors: acquisition of new natural habitats, coupled with proportional increases in human population, and gradual accumulation of ecosystem information. By today's standards, both factors grew at a snail's pace, though: according to Kremer (1993a) data, the population growth rate between 25 000 BCE and 5 000 BCE amounted to 0.002% per annum. The resultant population doubling period amounted to 34 360 years, so the hunter-gatherer population even failed to double in this period.

In the agricultural era, in contrast, new technological developments allowed to transform natural habitats, intensify food production, and hence increase population density. At the same time, knowledge accumulation also somewhat accelerated thanks to a few breakthrough ideas like writing or philosophy, and to the gradually increasing scale of operations in proto-"R&D". There was also a positive feedback loop with the hunter-gatherer economy in the sense that farming was instrumental in acquiring additional habitable land. However, as eventually all useful land was subjected to human control, a "secular stagnation" in populated land area started to set in. According to Kremer (1993a) and Piketty (2014) data, in the period between 5 000 BCE and 1500 CE the population growth rate amounted to 0.069% per annum, implying a doubling period of 1007 years. While still modest by modern standards, it amounted to a massive 34-fold increase in the population growth rate compared to the preceding hunter-gatherer era. Owing to the Malthusian mechanism, though, increases in human control per capita Q/POP were very limited. According to Piketty (2014) data, growth in GDP per capita (PPP) in 0–1500 CE amounted to just 0.012% per annum (so that it would potentially double only in 5728 years).

Gradual progress in science and technology during the agricultural era, with scattered acceleration episodes in ancient Greece, the Roman Empire, or Song and Ming China, eventually culminated in the European scientific revolution of the sixteenth century and beyond, with its printed books and scientific method. New technologies were complementary to the existing agricultural economy, allowing for the spread of the Western civilization and an overall acceleration in population growth. According to Piketty (2014) data, between 1500 and 1820 CE the world population was growing at an average rate of 0.271% per annum, implying a doubling period of 256 years. At the same time, establishment of global trade links, exploitation of natural resources and development of crafts led to a gradual acceleration in output per capita. GDP per capita (PPP) grew in 1500–1820 CE at a rate of 0.053% per annum, which was still much slower than population growth, but already a fourfold increase compared to the preceding agricultural period.

The industrial revolution brought a further acceleration in economic growth. Breakthrough industrial technologies set off the processes of physical and human capital accumulation, and caused increases in GDP per capita and (relatively widely distributed) standards of living. There was also a positive feedback loop with the agricultural economy: mechanization of agriculture and spread of fertilizers increased crop yields, while developments in medicine, hygiene and increases in standards of living dramatically reduced mortality, paving the way for a demographic explosion (Boucekkine et al., 2003). A few decades after, however, the children quantity–quality trade-off associated with growing demand for skills and rising educational attainment triggered a decline in fertility, gradually bringing population growth rates down. According to Piketty (2014) data, in the period 1820–1990 CE the headline industrial-era measure of development, world GDP (in PPP), grew at a rate of 2.14% per annum, doubling every 33 years. The intensive margin of growth, GDP per capita, grew at a rate of 1.17% per annum, doubling every 60 years. At the same time, global population grew at a rate of 0.96% per annum, doubling every 72 years. (Handling global population dynamics requires caution because the demographic transition affected various parts of the world at different times. Population growth has already fallen to zero or even mildly negative values in most developed countries, a clear "secular stagnation" in population growth. However, according to Piketty (2014) data, in 1990–2012, global population grew at an unprecedented rate of 1.3% per annum, implying a doubling period of 53 years. This aggregate figure was driven exclusively by population growth in largely agricultural societies of, e.g., Sub-Saharan Africa or Southeast Asia, though. But these countries are already seeing reductions in average fertility rates, too (Rosling, 2018). United Nations population projections expect a stabilization of the global population at around 11 billion in 2100.)

Growth acceleration brought by the digital revolution worked through two channels: faster knowledge accumulation and the opening of a new dimension of economic growth, useful data. Production of new knowledge was accelerated thanks to the development of digital R&D equipment, facilitating knowledge storage and retrieval, connecting the researchers via the Internet, and automating the most tedious research tasks. In terms of the new dimension of growth, in turn, the digital revolution started the processes of rapid accumulation of programmable hardware and its pre-programmed software. Furthermore, a strong feedback loop was activated with the industrial era via automation, increased quality and variety of goods, improved information flow that reduced firms' operating costs, and the opening of new production and service sectors. According to Hilbert and López (2011) data, between 1986 and 2007, processor capacity grew at 58% per annum (doubling every 1.5 years, the exact Moore's Law), data storage grew at 23% per annum (doubling every 3.3 years), and data communication grew at 28% per annum (doubling every 2.8 years). Owing to the positive feedback loop with the industrial-era economy, GDP growth seems to have also accelerated in the digital era. According to Piketty (2014) data, in 1990–2012 CE, the average GDP per capita (PPP) growth rate was equal to 2.08% (doubling every 34 years).

As an aside, note that although annualized growth rates and doubling times are a convenient descriptive tool, exponential growth in any particular dimension shouldn't be expected to continue indefinitely. From the theoretical angle, that would call for making highly unlikely "knife-edge" assumptions (Growiec, 2007). Also factually, looking at the millennia of past economic growth suggests that sigmoid (S-shaped) long-run development patterns are probably more accurate depictions of reality: habitat acquisition has been completed long ago, global population growth is projected to stop at the turn of the next century, and it is conceivable that global GDP will eventually stagnate, too.

Despite this caveat, the evidence for past growth accelerations is overwhelming. Not only have the consecutive technological revolutions accelerated development by orders of magnitude, but they have also been getting closer to one another in time (Table 7.5). It took our ancestors as much as 60 000 years (starting the clock at the cognitive revolution) to first initiate farming. It then took 9 500 years for farming to spread around the world and to lay the ground for the scientific revolution. From then on, the consecutive revolutions were much closer to each other in time: the industrial

Table 7.5 Estimates of timing and magnitude of growth accelerations due to technological revolutions

Revolution	Measure	Lag	Acceleration	Cumulative
Agricultural revolution	Population	60 000 y	34.1	34
Scientific revolution	World GDP	9 500 y	4.7	161
Industrial revolution	World GDP	250 y	6.6	1061
... *scientific+industrial*	*World GDP*	*9 750 y*	*31.1*	*1061*
Digital revolution	Data	180 y	13.1	13880

Sources Kremer (1993a), Piketty (2014), Hilbert and López (2011). See the details in text

revolution happened in about 250 years, followed by a digital revolution in about 180 years. Given that it's already been around 40 years since the digital revolution, it is quite conceivable that another breakthrough may occur within a couple of decades from now.

All in all, the four technological revolutions from 10 000 BP until today (and there are no reliable data to quantify the acceleration of growth following the cognitive revolution) have accelerated the pace of growth in world economic development by a staggering combined factor of 13 880. An almost 14 000-fold acceleration! And OK, if you dismissed data communication as a valid indicator of the pace of development in the digital era and instead used the conservative measure of world GDP, the combined factor would go down by an order of magnitude, to 1689. Nevertheless, either way you take these numbers, they clearly mark the humankind's breathtaking success in controlling the world.

7.5 Feedback Loops Across the Eras

What blurs the sharp distinction between the consecutive eras, however, is that they are interrelated. Each new era opens an entirely new dimension of development but also greatly affects the previous-era economy. Such positive feedback effects were visible across all eras.

Firstly, the cognitive revolution enabled the accumulation of ecosystem information. Thereby it created advantages of the *homo sapiens* against other species, facilitating the acquisition of new habitats under human control, Hab.

Secondly, the agricultural revolution allowed our ancestors to transform ecosystems and make them more habitable. Several previously hostile and thus sparsely inhabited biomes such as steppes have become fertile once domesticated crops and animals have been introduced. Accumulated experience in farming also helped our predecessors refine and improve their knowledge of the underlying biological features of crops and domesticated animals, which *inter alia* allowed them to harness the power of animal muscles to perform a range of physical tasks in their service.

Therefore the hunter-gatherer measures of economic development, habitat capacity Hab and ecosystem information EI, were both increasing in the agricultural era.

Thirdly, the scientific and industrial revolutions not only opened up a new (industrial) sector of the economy but also revolutionized the old one: farming. Mechanization of agriculture, nitrogen, phosphate and potassium fertilizers, and the Green Revolution contributed to massive increases in crop yields. At the same time, progress in the science of medicine implied a massive drop in mortality. Both effects led to a huge cumulative improvement in the agricultural-era measure of development, total population POP (and population density). However later fertility dropped, and nowadays rapid population growth is observed only in the poorest regions of the world where people are still employed in subsistence agriculture—and even there fertility is gradually falling with development.

Fourthly, industrial-era developments increased scientific productivity by providing ever more sophisticated laboratory equipment, culminating in such monuments of science as the Very Large Telescope (VLT) in the Chilean Atacama desert and the Large Hadron Collider (LHC) under Geneva. Scientific inquiry in the nineteenth and twentieth century, especially in natural and life sciences, has become very different to what it used to be during the Enlightenment, and scientific knowledge SK grew accordingly.

Fifthly, modern digital-era technologies massively improve the efficiency of industrial production and R&D. There are many direct channels: cost-reducing and quality-enhancing automation, increasing quality and variety of goods thanks to the featuring of electronics, improved information flow within and among firms, opening of new service sectors (business-to-business, ICT services, etc.), large-scale use of computing in science and engineering, etc. Furthermore, improved information flow and processing also indirectly improved industrial efficiency by facilitating globalization—in particular global fragmentation of production and the creation of global value chains. All that provided major boosts to global GDP. Robert Solow may have said in 1987: "You can see the computer age everywhere but in the productivity statistics" (Solow, 1987), but nobody would dare to say this today. Effects of the digital era have become clearly visible also in productivity statistics.

Impressive as they are, however, these are only *second-order effects* of a new era feeding back on the earlier eras. First-order effects, in contrast, are present directly in the dimension of development opened by the most recent era. In the current digital era, first-order effects of development are seen in the pace of accumulation and processing of useful data: Moore's law, explosion of the volume of Internet content and the outstanding progress in AI, all too often summarized by "who would've thought that...".

The flipside of massive growth accelerations caused by technological revolutions are "secular stagnations" which appear in the previous-era economy whenever the positive feedback loop from the next era has subsided and further development remains concentrated only in the most recent era. Nowadays the topic is hotly discussed in the context of GDP growth slowdowns observed across the developed economies since the 2000s s (e.g., Gordon (2016), Cette et al. (2016), Dong et al. (2016)). The debated "secular stagnation" hypothesis—that in the next decades GDP growth in developed economies will permanently slow down, eventually to zero—can be intuitively framed in the current framework as hypothesized arrival of a period when positive feedback effects from the digital-era economy on the industrial economy have subsided and further growth remains concentrated only in the digital era, with no subsequent technological revolutions in hardware.

Is this already happening? Certainly not. We are still at an early stage of the digital era and there is still ample room for development in AI algorithms, able to massively feed back on global GDP (cf. Brynjolfsson et al. (2019)). Will it happen in more distant future? Probably one day it will, but the timing of that event is completely unknown.

7.6 Inequality Between and Within the Eras

What *further* blurs the sharp distinction between the consecutive eras is that they may co-exist in time and space. This has profound implications for *inequality*, i.e., the variation within the distribution of certain variables across the human population, such as income, wealth and the consumption of certain goods and services. (I expect that in the digital era it may also make sense to compute inequality in access to useful data and computing power.) Global distribution of income is a huge topic which I consciously set aside in this book. But there are two breakdowns of inequality that are relevant here: factor shares vs. concentration of factor ownership, and within-era vs. between-era inequality.

The factor shares vs. concentration of factor ownership decomposition naturally corresponds with the hardware–software model of production and growth. The key logic is that changes in factor share automatically bring changes in the degree of income and wealth inequality as long as ownership of one factor is more concentrated than the other. For example, accumulation of capital in the early industrial era increased income inequality because capital tends to be more concentrated than labor (Galor, 2005). Human cognitive work and physical labor, note, are naturally dispersed by being embodied in humans. Later on, however, income inequality declined because of an increasing share of output being attributed to human cognitive work (skilled labor) at the expense of physical capital. On top of that concentration of capital in the Western economies went down in the early twentieth century (Piketty and Zucman, 2014), during the two world wars. In consequence, the time path of inequality broadly followed the pattern of the capital share: except for the decline in inequality around both world wars, it first went up and then down.

In the digital era for the first time in history the software factor includes not only human cognitive work but also disembodied pre-programmed software, whose ownership is highly concentrated. As a lion's share of the incremental software remuneration in the digital era goes to computer software owners rather than users, increases in the software share of output are now associated with increasing, not decreasing income inequality. This change is probably one of the key drivers of the landmark reversal of macroeconomic trends over the last 40 years: declining labor shares (Karabarbounis and Neiman, 2014), increasing profit shares (Barkai, 2020), and increasing income inequality, especially at the top of the distribution (Andrews et al., 2016; Jones and Kim, 2018; Piketty, 2014).

Such observations should be however interpreted with caution and certainly not mindlessly extrapolated into the future. First, they are based on data that conflate the industrial-era and digital-era economy, and thus not only illustrate phenomena specific for the digital era but also capture the shift between the eras. Second, technological revolutions also hugely affect relative prices. After the industrial revolution, there was a massive decline in the price of industrial goods relative to land and property. Similarly following the digital revolution we are observing a secular decline in the price of machinery and equipment (especially programmable digital hardware—ICT capital) relative to other goods (especially non-durable consumption goods, land and property), cf. Greenwood et al. (1997), Fernald (2015). Third, the hardware–software model predicts that with sufficient improvements in AI, software will begin to effectively scale up with the available programmable hardware. After reaching an inflection point, software should then cease to be a bottleneck of growth and consequently its output share should fall down. Output will then be increasingly appropriated by the owners of hardware.

To understand the within-era versus between-era inequality decomposition, observe that (i) technologies from subsequent eras coincide across the world and (ii) there are marked growth accelerations after each technological revolution. For example, according to the Africa Agriculture Status Report 2017, about 70% of population in Sub-Saharan Africa is still employed in subsistence agriculture. At the same time, it becomes a continent of contrasts whenever we see a clash of the agricultural, industrial and digital eras. The Western world, in comparison, has largely eliminated subsistence agriculture, and achieved affluent and relatively equal societies. Yet, even there one may easily observe differences in the pace of development between the relatively stagnant "traditional" industry (e.g., the US Rust Belt) and the booming digital economy (e.g., the Silicon Valley).

Since growth rates differ by orders of magnitude between eras, the timing of takeoff into another era determines the relative standing of nations, regions, economic sectors and people (Diamond, 1997). After a new era begins in some regions of the world while others are left behind, between-era inequality quickly overwhelms within-era inequality, giving rise to a "great divergence" in economic development and bimodal (or more generally, multimodal) distributions of e.g., country or regional GDP (Henderson and Russell, 2005; Quah, 1997). Within countries, measured increases in inequality, in particular top income inequality and CEO pay, also partly represent between-era inequality, i.e., the inequality between those who got on the train and those who missed it. Later on, however, technology diffusion and upward real convergence are powerful forces of equalization in income, standards of living, and—most broadly—satisfaction of human needs (Pinker, 2018; Rosling, 2018).

This is how accelerating growth has worked for the humankind in the past. But what does it teach us about the future?

References

Andrews, D., Criscuolo, C., & Gal, P. N. (2016). The global productivity slowdown, technology divergence and public policy: A firm level perspective (Working Party No. 1 on Macroeconomic and Structural Policy Analysis). OECD.

Barkai, S. (2020). Declining labor and capital shares. *Journal of Finance,75*, 2421–2463.

Boucekkine, R., de la Croix, D., & Licandro, O. (2003). Early mortality declines at the dawn of modern growth. *Scandinavian Journal of Economics,105*, 401–418.

Brynjolfsson, E., & Oh, J. H. (2012). The attention economy: Measuring the value of free digital services on the internet. In *Paper presented at 33rd International Conference on Information Systems*. MIT.

Brynjolfsson, E., Rock, D., & Syverson, C. (2019). Artificial intelligence and the modern productivity paradox: A clash of expectations and statistics. In A. Agrawal, J. S. Gans & A. Goldfarb (Eds.), *The economics of artificial intelligence: An agenda*. University of Chicago Press.

Cette, G., Lecat, R., & Ly-Marin, C. (2016). *Long-term growth and productivity projections in advanced countries* (Working Paper No. 617). Banque de France.

Diamond, J. (1997). *Guns, germs, and steel: The fates of human societies*. Norton & Company.

Dong, J., Li, W., Cao, Y., & Fang, J. (2016). How does technology and population progress relate? An empirical study of the last 10,000 years. *Technological Forecasting and Social Change,103*, 57–70.

Fernald, J. G. (2015). Productivity and potential output before, during, and after the great recession. *NBER Macroeconomics Annual,29*, 1–51.

Galor, O. (2005). From stagnation to growth: unified growth theory. In P. Aghion & S. N. Durlauf (Eds.), *Handbook of economic growth* (pp. 171–293).

Galor, O. (2011). *Unified growth theory*. Princeton University Press.

Galor, O., & Moav, O. (2002). Natural selection and the origin of economic growth. *Quarterly Journal of Economics,117*, 1133–1191.

Gordon, R. J. (2016). *The rise and fall of American growth: The U.S. standard of living since the civil war*. Princeton University Press.

Greenwood, J., Hercowitz, Z., & Krusell, P. (1997). Long-run implications of investment-specific technological change. *American Economic Review,87*, 342–362.

Growiec, J. (2007). Beyond the linearity critique: The knife-edge assumption of steady-state growth. *Economic Theory,31*, 489–499.

Harari, Y. N. (2014). *Sapiens: A brief history of humankind*. Vintage.

Harris, S. (2010). *The moral landscape: How science can determine human values*. Free Press.

Henderson, D. J., & Russell, R. R. (2005). Human capital and convergence: A production-frontier approach. *International Economic Review,46*, 1167–1205.

Hilbert, M., & López, P. (2011). The world's technological capacity to store, communicate, and compute information. *Science,332*, 60–65.

Jones, C. I., & Kim, J. (2018). A Schumpeterian model of top income inequality. *Journal of Political Economy,126*, 1785–1826.

Karabarbounis, L., & Neiman, B. (2014). The global decline of the labor share. *Quarterly Journal of Economics,129*(1), 61–103.

Kremer, M. (1993a). Population growth and technological change: One million B.C. to 1990. *Quarterly Journal of Economics,108*, 681–716.

Maslow, A. (1954). *Motivation and personality*. Harper & Brothers.

Mazzucato, M. (2018). *The value of everything: Making and taking in the global economy*. Public Affairs Books.

Piketty, T. (2014). *Capital in the twenty-first century*. Translated from French by Arthur Goldhammer. Harvard University Press.

Piketty, T., & Zucman, G. (2014). Capital is back: Wealth-income ratios in rich countries, 1700–2010. *Quarterly Journal of Economics,129*(3), 1255–1310.

Pinker, S. (2018). *Enlightenment now: The case for reason, science, humanism, and progress*. Viking.
Quah, D. T. (1997). Empirics for growth and distribution: Stratification, polarization, and convergence clubs. *Journal of Economic Growth,2*, 27–59.
Romer, P. M. (1990). Endogenous technological change. *Journal of Political Economy,98*, S71–S102.
Rosling, H. (2018). *Factfulness: ten reasons we're wrong about the worldan—d why things are better than you think*. Sceptre.
Solow, R. M. (1987). We'd better watch out. *New York Times Book Review,36*.
Stiglitz, J., Sen, A., & Fitoussi, J.-P. (2009). *Report by the commission on the measurement of economic performance and social progress*. Sorbonne.

Chapter 8
Accelerating Growth in the Future

We won't experience 100 years of progress in the 21st century—it will be more like 20,000 years of progress (at today's rate). (...) There's even exponential growth in the rate of exponential growth.

Ray Kurzweil

There was virtually no growth before 1750, and thus there is no guarantee that growth will continue indefinitely. (...) The rapid progress made over the past 250 years could well turn out to be a unique episode in human history.

Robert Gordon

As Niels Bohr used to say, "it is very difficult to predict, especially the future". Secular trends are best identified *ex post*, and even then they are often challenged and revised. I have two aims in this chapter. First, to specify the data that would be most helpful for improving our foresight into the digital era, making the grounds for our long-run growth predictions as firm as possible. Second, to use the imperfect evidence that already exists and place my bets for the coming decades against two polar benchmarks: secular stagnation and technological singularity.

8.1 Global Data Accounts

If we agree that the digital era is able to increase the extent of human control and the degree of satisfaction of human needs beyond the pecuniary value of its goods and services, the implication is that our existing aggregative measures of economic

J. Growiec, *Accelerating Economic Growth*, Frontiers in Economic History,
https://doi.org/10.1007/978-3-031-07195-9_8

development such as GDP paint a very incomplete picture of the present. The digital revolution has, note, for the first time in history decoupled total data collection and transmission from GDP, and data processing—from human brains.

The concept of GDP, with its focus on the *dollar value* of inputs and outputs of production processes, is historically rooted in the industrial era. It ignores the fact that differences in information content per unit of value added, unacknowledged in the national accounts, may carry economic meaning. For instance, several classes of digital goods, such as online entertainment (free music, movies, etc.), open source and freeware utility software, or open access scholarly publishing, are entirely missed by the national accounts even though they contribute to the well-being of the society (Brynjolfsson and McAfee, 2014), satisfaction of human needs, and ultimately the aggregate span of human control.

Therefore, wishing to capture the welfare gains from the tremendous growth in the world's cumulative ability to compute, store and communicate useful information, far exceeding GDP growth rates in the last decades, we should think about devising a digital-era counterpart of the national accounts, capturing the flows of *useful data* between people and autonomous digital devices. This is now technologically feasible: in contrast to earlier eras when information transmisson was done exclusively through conversations among people (in person or paper), from the 1980s onward, it is carried out increasingly through digital means, leaving digital traces that could be quantified.

More specifically, to account for the intensive margin of development in the digital era we need a concept that would work like value added in the national accounts. In particular, *useful* data should be distinguished from *all* data, perhaps in a manner similar to the distinction between value added and global output. An important difference is, though, that information is non-rivalrous, so that the source dataset can be costlessly (and typically is) retained after transmission. One should also redefine the sectoral breakdown of the economy in order to get a more precise measurement of the structure of data flows. In my perception, devising such new measurement methods will be as desirable as it is challenging. Preliminary ideas on how this could be accomplished have been proposed by information theorists (Hilbert, 2017; Hilbert and López, 2011), but many important questions remain open.

As a side benefit, setting up a system of data accounts may also indirectly facilitate the measurement of technological knowledge. Thus far the latter variable has proven to be inherently difficult to measure. The best proxies economists use in this regard are total factor productivity (TFP), patents and research articles plus scientific books. They are all very crude measures of technological knowledge, though. TFP is a residual macroeconomic measure, which also includes all sorts of measurement and specification error. Direct measures like patents or research articles and books, in turn, are problematic because of incomplete treatment of actual usefulness and information content of each patent, article or book. Detailed data accounts could potentially rectify this problem.

There are many caveats, though. First, data flows rarely respect national boundaries, so that the measurement ought to be done at the global scale. Therefore, I am in fact calling for Global (not National) Data Accounts (GDA). Second, there is a lot of redundancy in data transmission, storage, and even creation. A lot of thought must be devoted to the filtering and compression of raw data flows so as to distill the key quantities in question. Third, some data flows may be unintended or detrimental to the users—just think, for example, about the spread of fake news, Internet hate, and computer viruses. Fourth, the data accounts would have to cleverly embrace the evolving role of intellectual property and find a clever way around data secrecy in order to adequately compute the volume of data flows without compromising the proprietary character of some datasets.

Of course, this is not to say that the System of National Accounts should be discontinued. To the contrary, in fact data accounts would only *improve* the accuracy and usefulness of national accounts. The digital revolution has opened up a new dimension of development, which hasn't been sufficiently measured yet. So far we could only indirectly evaluate the effects of the digital-era economy by accounting for the feedback effects from the digital era to the industrial era, present in the data on GDP and value added (e.g., Jorgenson and Stiroh (2000), Timmer and van Ark (2005)). But we are still failing at capturing total effects of digital development.

It is also critical to distinguish between inputs and outputs of the digital-era production process. Flows of useful data—an output variable—are generated using the previously accumulated inputs: stocks of hardware and software, both of which can and should be measured, too.

Despite advocating for Global Data Accounts and arguing that flows of useful data are a viable measure of the state of development of the human civilization, I don't see any reason to believe that the accumulation, transmission or processing of useful data in itself *is* local control. Saying so would amount to adopting *dataism*, a worldview that "declares that the universe consists of data flows, and the value of any phenomenon or entity is determined by its contribution to data processing" (Harari, 2017, p. 428). In contrast, in my view, collecting and processing useful data are not the ends, but means to obtaining more control. There are two key differences between my standpoint and the dataist one. First, I think that exercising control over the environment must include, to a non-negligible extent, carrying out physical actions—serving at the very least to produce material outputs that feed our brains and sustain our material hardware. These actions must be consistent with the laws of thermodynamics, and thus, in particular, require expending energy. To my knowledge, such actions can't be reduced to information processing. (Unless physical reality is an illusion and we're living in a simulation, Bostrom (2003).) Second, instead of sticking to the volume of useful data (or perhaps some other measure that would feature information volume, complexity or entropy) as the ultimate measure of development of humankind I am open to the possibility that another technological revolution would come along in the future, opening yet another dimension of activity and justifying a split of the level of development Q into further "post-digital" factors. In fact, there are already a few possible candidates for such breakthrough technologies (Betz, 2018), such as synthetic biology, nanotechnology and applications of quantum physics.

8.2 Inputs and Outputs of the Digital Era

Following my call to design and implement Global Data Accounts in order to properly capture the digital-era output variable, *flows of useful data*, I would also like to call for better measurement of the new inputs of the digital-era production process, *programmable hardware* and *pre-programmed software*. It would be also good to learn how to measure the hardware and software share of digital output. Yet another challenge for measurement is digital inequality, that is: (i) inequality in people's access to computing power, data storage capacity and bandwidth—which are stock variables and so this concept is akin to wealth inequality and (ii) inequality in the in- and outflows of useful data from their digital devices—which are flow variables, making this idea similar to income inequality.

Among the two inputs of digital production, measurement of hardware is probably going to be relatively easier. We can reasonably expect that the key characteristics of "programmable hardware", including all capital and consumption goods, which have a digital component that is able to store and execute code, are likely going to be computing power, data storage capacity and bandwidth (Hilbert and López, 2011). The software factor is going to be way more challenging, though. The aggregate volume and/or complexity of (somehow compressed and normalized) instructions provided to hardware should include both the flows of services of existing pre-programmed software, AI, and the cognitive work of human programmers and other users. A caveat is that digital software can be costlessly copied and run simultaneously on many devices.

A related challenge is to quantify hardware *capacity utilization*. My working hypothesis in this regard is that capacity utilization of computing power and bandwidth is generally low nowadays, much lower than overall physical capital utilization. The computing power and bandwidth of our computers are frequently left idle because our software—and particularly so, the human users—provide the hardware with relatively very few instructions. Therefore huge returns are still waiting for the developers of new software (e.g., AI algorithms) which could effectively tap this unused potential and generate large increases in the extent of human control.

There are also potentially interesting distributive implications from quantifying the volume of idle computing power and digital memory. Namely, their substantial fraction is probably going to be embodied in personal computers and other electronic devices whose ownership is dispersed. This raises a question if we could potentially implement a "computer banking" system allowing users to rent their idle computational capacity and digital memory, so that they could use their electronic devices as a source of capital income. That would act to reduce income inequality. On the software side, however, such institution would probably further reinforce the development of sophisticated software replacing human work and thus indirectly increasing inequality. Thinking of striking a balance between both forces we have to bear in mind, though, that the latter phenomenon is likely to continue into the future anyway, regardless of what we do with our electronic devices in the meantime.

I also expect that there will be less and less scope for comparing the "wealth of nations" in the fully globalized digital era. The primary level of measurement should be therefore gradually shifted from the country level to the global value chain (that is, value network) level and the global level. In fact, world's top technological companies are already multi-national and almost footloose. For example, according to Eurostat data real GDP (PPP) of Ireland grew in 2015 by a staggering 25.6%, against the European Union average of 2.3%. Why? Because in 2015, a couple of large multi-national but hitherto US-based companies moved their productive assets to Ireland and/or domiciled in Ireland by buying a smaller Irish-registered rival and "inverting" into an Irish corporate structure. (For example, Apple moved to Ireland its copyrights and patents on the design and technology.) They did so to save on taxes. It is however doubtful if this statistical fact had any bearing on the actual incomes, productivity and local control of Irish citizens.

8.3 Evidence of a Data Explosion

The key reason why we need Global Data Accounts is that they would give us a handle on the ongoing development in the data sphere and could take a central position in informing the global digital policy, the need for which I already advocated above. Frankly, Global Data Accounts could do the same job for digital policy as national accounts are doing for our contemporary, largely industrial-era and country-level economic policy.

Data measurement is clearly not a futile task and first steps in that direction have already been made, even if at a crude level that is not yet sufficient for economic policy purposes. Namely, information theorists have produced their first assessments of the level and growth of the world's total storage, transmission and computation of data over the last decades. The seminal contribution of Hilbert and López (2011) tracks 60 analog and digital technologies during 1986–2007 and announces that: "In 2007, humankind was able to store 2.9×10^{20} optimally compressed bytes, communicate almost 2×10^{21} bytes, and carry out 6.4×10^{18} instructions per second on general-purpose computers. General-purpose computing capacity grew at an annual rate of 58%. The world's capacity for bidirectional telecommunication grew at 28% per year, closely followed by the increase in globally stored information (23%)."

To put these numbers in perspective, digital storage, transmission and computation of data can be compared with their carbon-based counterparts: memorizing, communicating and thinking that we have been doing for millennia using our human brains. (Before the invention of writing, human brains were the only store of data. After writing, and especially the printing press, data storage was gradually outsourced to "external memory"—books, journals, letters—allowing the stock of stored data per person to rise.) The digital revolution has been a real game changer in this regard. Gillings et al. (2016) write: "Evolution has transformed life through key innovations in information storage and replication, including RNA, DNA, multicellularity, and culture and language. We argue that the carbon-based biosphere has generated a cognitive system (humans) capable of creating technology that will result in a comparable evolutionary transition. Digital information has reached a similar magnitude to information in the biosphere". Most of this explosion took place in the last three decades as the world's capacity to store, communicate, and compute information has begun to soar only since the 1980s (Hilbert and López, 2011). Quantitatively, "[t]he digitalization of the entire stockpile of technologically-mediated information has taken less than 30 years. Less than one percent of information was in digital format in the mid-1980s, growing to more than 99% today. (...) Information technology has vastly exceeded the cognitive capacity of any single human being (...). In terms of capacity, there are two measures of importance, the number of operations a system can perform, and the amount of information that can be stored. The number of synaptic operations per second in a human brain has been estimated to lie between 10^{15} and 10^{17} (...). While this number is impressive, even in 2007, humanity's general purpose computers were capable of performing well over 1×10^{18} instructions per second (...). Estimates suggest that the storage capacity of an individual human brain is about 10^{12} bytes (...). On a per capita basis, this is matched by current digital storage (5×10^{21} bytes per 7.2×10^{9} people)" (Gillings et al., 2016).

Two observations are due here. First, these estimates, while certainly impressive, are exact only up to an order of magnitude. This is not yet the statistical precision we are accustomed to in economics and policy.

Second, these estimates include a mixture of input and output, stock and flow variables. Data storage capacity and processor computational capacity can be thought of as stock input variables, akin to physical capital used in industrial production. The amount of actually communicated data is, on the other hand, a flow output variable. Unfortunately, it's not yet the postulated concept of *flows of useful data*. To reach that level of precision, one would have to filter out the data redundancy as well as intermediate, auxiliary data used up in computations. In economic terms, total communicated data is a digital-era equivalent of global production, and what we need is flows of useful data—an equivalent of value added.

8.4 Singularity Is Near?

As documented above, between the cognitive revolution of 70 000 BP and today, growth in human control accelerated by a staggering combined factor of 13 880. Moreover, not only has each of the past five technological revolutions accelerated development by at least an order of magnitude but also the consecutive revolutions were getting ever closer to each other in time (Table 7.5). What are then the perspectives for the future? Should we expect a cascade of further technological revolutions culminating in a technological *singularity* around 2050 CE (Kurzweil, 2005), a secular stagnation in economic and technological development (Gordon, 2016), or perhaps something in between?

Predicting the future without a crystal ball requires making good assumptions. In my case this means following the theory laid out in this book. My expectations are as follows. (A similar thought experiment was recently carried out by the Open Philanthropy foundation, cf. Davidson (2021).)

First, I expect an eventual decline in the positive feedback effects from the digital era to the industrial era, leading to a "secular stagnation" first in total population (probably around 2100 CE, according to United Nations projections) and later in world GDP. A secular stagnation in world GDP should happen no earlier than in the twenty-second century because global population growth needs to stop first, and industrial technologies need to fully diffuse around the world, so that the relatively backward economies can largely catch up with the world leaders in GDP per capita. Despite both upcoming slowdowns, I expect to see a continuation of the soaring growth rates of information creation, storage and communication that we've been observing since the 1980s. Therefore, in the coming decades, aggregate human control should continue to grow about an order of magnitude faster than world GDP. This is my expectation conditional on the assumption of no global disaster such as a nuclear world war or runaway global warming, and no further technological revolution in hardware.

My second expectation is, however, that a further technological revolution in hardware is actually quite likely to happen within the twenty-first century, perhaps even before 2050. Extrapolating into the future the observation that consecutive revolutions accelerate development by an order of magnitude and get closer to each other in time, Hanson (2000) speculates—and I agree—that "within the next century a new mode might appear with a doubling time measured in days, not years". This would be my expectation conditional on the assumption that we will continue to see similar rates of progress in development in computer software, and especially AI, as we've been observing since the 1980s.

But why would the likelihood of a next technological revolution in hardware depend on the progress in software development? The answer is simple: in order to develop new ways of performing physical actions which would radically increase our capacity to pursue local control, we need to acquire the requisite knowledge first.

Thus far, the software factor is the key developmental bottleneck of the human civilization. Instructions provided to our hardware, both in production and R&D, are still largely based on human cognitive work, which is a problem because while the stock of programmable hardware is growing at the pace of Moore's law, doubling every 2–3 years, human cognitive powers are too limited to keep up (and even sustain any exponential rate of growth). The incentives to develop advanced AI are therefore increasing fast. If we were to expect a future without superhuman, self-improving and self-replicating AI, then the discrepancy between hardware capabilities and software limitations would continue to grow and eventually become arbitrarily large. After all, by 2100 CE, the 11 billion unaugmented human brains, even if fantastically educated, will only be able to process as much information per hour as is allowed by their biological structure. And we are, arguably, "pulling ourselves by our bootstraps" for quite a long time already. Our collective brain is continuously augmented with an increasing frequency of teamwork in business and research, allowing to pool our cognitive resources when solving increasingly complex issues. We fight against the exponentially growing knowledge base by increasing average educational attainment, investing in life-long learning, and narrowing down our expertise. We increasingly outsource our memory and computational capacity to external devices, and rely on the Internet with its AI-based search engines for finding the relevant information. There are no "Renaissance men" in the world anymore, and communication of research results to the general public is becoming increasingly difficult—not because people are getting dumber, but because the science is getting more and more complicated. All in all, current rates of exponential growth in information collection and processing can be maintained over the next decades, thus leading to a high probability of a next technological revolution, only if we successfully outsource our cognitive tasks to rapidly developing AI software whose cognitive powers will be able to keep up with the progress in hardware.

Accordingly, my third expectation is that within the current century, and potentially even before 2050, one of our AI labs will actually produce artificial superintelligence. (The perils of which I will cover in a separate chapter.) Of course I'm not willing to bet any money if that's going to happen before 2050 or not, but I'm quite confident that one day it is going to happen, following our drives for creativity and technological perfection, instrumental in maximizing local control. However, if it *never* happens and AI development hits a solid wall, the chances for a next hardware revolution will go down. In that case, please return to my expectations from point one above.

So is singularity near? Let's check what that means. There are at least three alternative definitions of technological singularity in the literature: (i) "a future period during which the pace of technological change will be so rapid, its impact so deep, that human life will be irreversibly transformed" (Kurzweil, 2005), (ii) the arrival of artificial general intelligence (AGI) with superhuman abilities (Bostrom, 2014), and (iii) a vertical asymptote in development, i.e., a situation where a certain variable (e.g., GDP, digitally stored information, aggregate development Q) tends to infinity in finite time (Nordhaus, 2021). These definitions have a lot in common but aren't equivalent.

Ray Kurzweil's definition of singularity is both the most intuitive and most general of all three, even if somewhat tautologically stated (wasn't human life irreversibly transformed already many times in the past, e.g., by the arrival of sedentary agriculture or flush toilets?). It resonates well with our perception that modern days may be crazy, but don't quite qualify as singularity—whereas for example a world where global output doubles every month would certainly qualify (could you then make any plans for summer?). Minding the cognitive constraints of the human brain, however, it is hard to imagine such a scenario happening without an operating AGI, or perhaps something similar to that, like vastly AI-augmented human minds or an ecosystem of interconnected artificial intelligences.

While Kurzweilian singularity doesn't necessarily imply the existence of AGI, the converse is certainly true: the existence of AGI implies Kurzweilian singularity. This is because from the instrumental convergence thesis we know that the AGI, whatever its goal is, will also maximize local control, exhibiting in particular drives for technological perfection and creativity. Factoring in the superhuman capabilities of that AGI, certainly "the pace of technological change will be so rapid, its impact so deep, that human life will be irreversibly transformed".

The vertical asymptote definition of singularity, in turn, is the strictest one: a scenario where global output doubles every month doesn't qualify as singularity anymore. On the one hand, this definition is impossible to satisfy literally: because a positive fraction of output must be material to sustain the hardware, such singularity is inconsistent with the laws of thermodynamics. On the other hand, it is nevertheless a good approximation of a scenario where any physically conceivable level of technological advancement can attained virtually instantaneously. (In this regard, estimating a super-exponential growth pattern with power-law acceleration in the growth rate, Johansen and Sornette (2001) find that the data on global GDP are consistent with a spontaneous singularity in the year 2052 ± 10, "signaling an abrupt transition to a new regime". Similar super-exponential growth patterns have been documented in ICT data (Nagy et al., 2011). It remains an open question, however, if this seeming infinity isn't only an artifact of poor approximation of sharply rising but finite-valued time paths with hyperbolic curves which possess a vertical asymptote.)

Summing up, it seems that singularity—especially under the broad Ray Kurzweil's definition—is not yet near, but likely coming. The key variable that will determine the likelihood and timing of an upcoming singularity is the progress in AI capabilities. (Achieving an arbitrarily high level of software development in finite time is infeasible, but crossing the human level of intelligence is not as far-fetched. After all, as Stephen Hawking said, "there is no physical law precluding particles from being organized in ways that perform even more advanced computations than the arrangements of particles in human brains".) Further questions are then, will the actions of superhuman AI produce a next technological revolution, and will they continue to benefit human control?

8.5 Secular Stagnation? No, Thank You!

The economic mainstream is highly skeptical about the "singularitarian" point of view. By stark contrast, one of the major worries of modern day macroeconomists is that we may be approaching a *secular stagnation*, permanent decline in the pace of economic growth. Although mutually contradictory, both predictions are actually well grounded in data. The difference is that singularitarian techno-visionaries like Ray Kurzweil prefer to look at the frontier of technological capabilities of humankind and observe its evolution at long time scales, whereas down-to-earth "stagnationist" economists like Robert Gordon prefer to look at actually realized GDP per worker in a perspective of, at most, decades.

To illustrate the contrast, Gordon (2016) names three radical innovations which speeded up technological progress since the eighteenth century: (i) the inventions of steam engine, railroads, and cotton spinning (1750–1830), (ii) electricity, the internal combustion engine, and running water with indoor plumbing (1870–1900) and (iii) ICT technologies: computers, Internet, and mobile phones (since 1960). According to his assessment, the economic effects of the first two radical innovations lasted about a hundred years but those of the third one are probably going to be more short-lived and are already phasing out. He names six "headwinds" to US economic growth in the near future: demography, education, inequality, globalization, energy/environment and the overhang of consumer and government debt. Then he proceeds to predict that GDP growth rates observed in two decades before the economic recession of 2007–2009 will never come back.

I think it's not at all where this world is heading. With all respect to Robert Gordon's work, I think that he (and plenty of other economists persuaded by similar arguments) may be misled on this issue by failing to acknowledge that digital technologies—programmable hardware, pre-programmed software and AI—affect, in the first place, not our ability to produce goods and services, but to store, communicate and process information. As a second-order effect, they *do* improve labor productivity, but their first-order effect is intangible—and therefore largely not included in the national accounts. But even when focusing on GDP, one shouldn't underestimate this intangible direct effect, though, for two reasons: first, its tangible productivity effects may come with a lag (Brynjolfsson et al., 2019) and second, digital technologies may eventually set the stage for a further technological revolution in hardware. In this perspective, evaluating the effects of ICTs focusing narrowly on short-run changes in GDP is somewhat like looking at the remarkable developments of 17th century physics and asking, "but how does it help feed more people?". (Yes, the science of physics *does* help feed more people after all, but it was probably not exactly what Isaac Newton had on his mind when writing his *Principia*.)

The bottom line is that long-run evidence, starting with the dawn of the *homo sapiens*, suggests unambiguously that we live in a world of accelerating, not decelerating change. Is this change going to lose momentum in the coming decades and lead us to a plateau of human development? It's not impossible but I really don't think so. Let's evaluate the scenarios which could lead to such an outcome.

First, building superhuman AI may turn out prohibitively difficult. In such a scenario, combined cognitive powers of the humankind armed with all the computing power of machines and devices available at any given point in time, just won't crack the problem of exceeding the human level of intelligence. The collective human brain would then, one day, hit the upper bound of its cognitive capacity, unable to process information any faster—and that would naturally constrain the pace of further technological progress. Nevertheless, even without any further hardware revolution such as nanotechnology or bio-engineering, in this scenario technological progress may still go on for centuries at a rate comparable or even faster than today, fueled by the accumulation of programmable hardware in R&D (Growiec, 2022). This progress would be accompanied by balanced, exponential economic growth.

So more restrictions are needed for the secular stagnation scenario to materialize. One of such constraints would be to assume—in line with some empirical evidence—that ideas are getting systematically harder to find (Bloom et al., 2020), and decreasing returns in R&D are actually sufficiently strong to effectively stop technological progress once the ideas are "fished out", regardless of the amount of resources thrown at the R&D sector. An alternative restriction would be to assume that one day the world economy will hit an upper bound of energy efficiency, and from then onwards, output generation will be critically constrained by the energy component of our hardware and no technological breakthroughs will ever alleviate this problem.

In sum, the secular stagnation scenario requires us not only to assume that human cognitive work will always be an essential factor in production and R&D but also to place a firm upper bound on the level of hardware or software in the aggregate production function and preclude further technological revolutions that could remove it. To me, this sounds unrealistically prohibitive.

Much worse than a secular stagnation, though, are scenarios of man-made disaster, such as a major nuclear war followed by a nuclear winter, runaway climate change, unstoppable engineered pandemics, or depletion of essential raw materials. (Natural disasters such as supervolcano eruptions or asteroid collisions are probably much less of a threat, Bostrom (2002).) Were any of these disasters to materialize, it would clearly have the potential to severely (permanently?) curtail the humankind's development potential. I think we should be worried about that because even if no particular threat is really likely, the stakes remain dizzyingly high. Albert Einstein once said: "I know not with what weapons World War III will be fought, but World War IV will be fought with sticks and stones". Yuval Noah Harari commented, in turn: "one thing that history teaches us is that we should never underestimate human stupidity". In response, Elon Musk comes with a backup plan to build a self-sustaining base on Mars: "If we were a multiplanetary species, that would reduce the possibility of some single event, man-made or natural, taking out civilization (...) It's insurance of life as we know it". In my opinion, over the long run insurance of life as we know it is much more important than fending off the risk of secular stagnation.

So yes, there are true risks of a long-standing economic stagnation or even decline. In my subjective evaluation, however, the probability that any of these risks materializes to an extent that would stop all further development is low. My choice is, therefore, to place a low weight on such scenarios.

Which doesn't quite make me an optimist; an "anxious optimist" will probably be a better term. My standpoint is: *worry not that development will stop, worry that development will be so great that it will spin out of control and accidentally hurt us.*

Throughout the human history, we have been breathtakingly successful in navigating around the consecutive filters and clearing the respective hurdles. But with increasing opportunities came increasing risks. The threats to the human civilization are nowadays as global and interdependent as our economy is. Perhaps for that reason we find it notoriously hard to address them in a socially optimal way. It's much easier to close our eyes and pretend they don't exist. (Have you seen the movie *Don't Look Up*?) For example, economic analysis generally disregards catastrophic risks. We act as if balanced growth at 2–3% per annum was always going to be "the normal" and the perspective of a secular stagnation—so ghastly an outcome that it must be avoided at all costs. We find ourselves unable to appreciate the scenario of permanent stagnation with no existential risks against the backdrop of a scenario where growth is fast but most likely ends in our extinction. It's a pity that due to our human craving for ever increasing local control, the perspective of a stagnant but affluent and sustainable world economy is rejected as unsatisfactory. It is, after all, a very safe outcome that minimizes the risk that we all die off—which, under business as usual, is predicted by participants of the Global Catastrophic Risk Conference in Oxford in 2008 (Sandberg and Bostrom, 2008) to happen before 2100 with a staggering 19% probability.

The bottom line is that our drive to maximize local control has placed our civilization on a knife edge where ever greater extents of our control are accompanied by a small but firm probability of disaster. And while this probability remains consistently low, the size of the foreseeable disaster is systematically growing with the advancement of technology. The first time in history when our ingenuity created an existential risk for the entire humankind was the development of nuclear weapons during World War II. In the future, however, nanotechnology, advanced bio-engineering and especially high-level AI will elevate the stakes even higher.

A final point is that thinking about the future one should carefully delineate between the consequences of considered scenarios for the humankind (the collective human brain) and for our civilization and the global economy. In a world where an increasing share of output is generated beyond human oversight and the probability of an emergent AGI is steadily increasing, outcomes on these two axes may no longer coincide. There may also be a situation where the world economy grows without humans participating. Possible long-run scenarios for our civilization will be somewhere along the spectrum: civilization collapse—secular stagnation—balanced or accelerating growth—singularity. Possible long-run scenarios for our species, in turn, will span the scale: extinction—stable control—increasing control—immortality. What should we expect? For the world economy I am quite confident that it's heading towards further growth accelerations and perhaps a technological singularity. For the humankind, however, my prediction is much more shaky. My sense of insecurity here comes from the fact that machine superintelligence, once set in motion will be impossible to stop (Hanson and Yudkowsky, 2013). Thus, if improperly designed, it will get out of our hand for the same reasons as human local control got of evolution's hand: pace (faster by orders of magnitude than the pace of human thought), recursive self-improvement, and greed for resources. If we succeed in building a friendly AGI serving human needs and always respecting our control, then "fortune favors the brave" and immortality is just around the corner. If we fail then the game is over and there is no respawn.

References

Betz, U. A. K. (2018). Is the force awakening? *Technological Forecasting and Social Change,128*, 296–303.

Bloom, N., Jones, C. I., Van Reenen, J., & Webb, M. (2020). Are ideas getting harder to find? *American Economic Review,110*(4), 1104–1144.

Bostrom, N. (2002). Existential risks: Analyzing human extinction scenarios and related hazards. *Journal of Evolution and Technology,9*.

Bostrom, N. (2003). Are you living in a computer simulation? *Philosophical Quarterly,53*, 243–255.

Bostrom, N. (2014). *Superintelligence: paths, dangers, strategies*. Oxford University Press.

Brynjolfsson, E., & McAfee, A. (2014). *The second machine age: Work, progress, and prosperity in a time of brilliant technologies*. Norton & Co.

Brynjolfsson, E., Rock, D., & Syverson, C. (2019). Artificial intelligence and the modern productivity paradox: A clash of expectations and statistics. In A. Agrawal, J. S. Gans, A. Goldfarb (Eds.), *The economics of artificial intelligence: An agenda*. University of Chicago Press.

Davidson, T. (2021). *Could advanced AI drive explosive economic growth?* Open Philanthropy. Report.

Gillings, M. R., Hilbert, M., & Kemp, D. J. (2016). Information in the biosphere: Biological and digital worlds. *Trends in Ecology and Evolution,31*, 180–189.

Gordon, R. J. (2016). *The rise and fall of American growth: The U.S. standard of living since the civil war*. Princeton University Press.

Growiec, J. (2022). What will drive global economic growth in the digital age? *Studies in Nonlinear Dynamics and Econometrics*, forthcoming.

Hanson, R. (2000). *Long-term growth as a sequence of exponential modes* (Working paper). George Mason University.

Hanson, R., & Yudkowsky, E. (2013). *The Hanson-Yudkowsky AI-Foom Debate*. Machine Intelligence Research Institute.

Harari, Y. N. (2017). *Homo deus: A brief history of tomorrow*. Vintage.

Hilbert, M. (2017). The more you know, the more you can grow: An information theoretic approach to growth in the information age. *Entropy,19*, 82.

Hilbert, M., & López, P. (2011). The world's technological capacity to store, communicate, and compute information. *Science,332*, 60–65.

Johansen, A., & Sornette, D. (2001). Finite-time singularity in the dynamics of the world population, economic and financial indices. *Physica A,294*, 465–502.

Jorgenson, D. W., & Stiroh, K. J. (2000). Raising the speed limit: U.S. economic growth in the information age. *Brookings Papers on Economic Activity,1*, 125–211.

Kurzweil, R. (2005). *The singularity is near*. Penguin.

Nagy, B., Farmer, J. D., Trancik, J. E., & Gonzales, J. P. (2011). Superexponential long-term trends in information technology. *Technological Forecasting and Social Change,78*, 1356–1364.

Nordhaus, W. D. (2021). Are we approaching an economic singularity? Information technology and the future of economic growth. *American Economic Journal: Macroeconomics,13*, 299–332.

Sandberg, A., & Bostrom, N. (2008). *Global catastrophic risks survey* (Technical report #2008-1). Future of Humanity Institute. Oxford University.

Timmer, M. P., & van Ark, B. (2005). IT in the European union: A driver of productivity divergence? *Oxford Economic Papers,57*(4), 693–716.

Part IV
Lessons for the Digital Era

Chapter 9
Side Effects of Growth

We're gonna win the Darwin Award, collectively.

Elon Musk, in conversation with Tim Urban of waitbutwhy.com

9.1 Local Control Implies Side Effects

Having discussed the key trends underlying the humankind's breathtaking success in controlling the world, and peeked into the future with the powerful tool of the almighty Trend Extrapolation, I'm ready to talk about the most important lessons from the analysis. But before you read on, a warning sign. Part IV of this book is focused on the future and therefore contains a fair amount of speculation. I do my best to make my guesses "educated", but they are still guesses, and as such they may go horribly wrong. *Caveat lector.*

One important lesson from past evidence is that accelerating economic growth, fueled by the unstoppable human drive to maximize local control, tends to have mounting unintended side effects. This is not surprising: even without *homo sapiens*, changes in natural ecosystems often have consequences reaching beyond the direct objectives which motivated these changes. By the second law of thermodynamics, if entropy is purposefully reduced in one place, it must increase somewhere else so that overall entropy of the universe doesn't go down. For example, when beavers dam a creek, lives of fish, reptiles or other mammals are also affected. Nature is a highly complex system, and humans—like all other existing intelligence—cannot fully comprehend its complexity. To this day we can only build highly simplified models of reality, and even with the best of our intentions, there are always aspects of reality that are not accounted for.

We do not pursue the best of our intentions, though. We don't even try to maximize long-run social welfare, and we're rarely preoccupied with sustainability or

J. Growiec, *Accelerating Economic Growth*, Frontiers in Economic History,
https://doi.org/10.1007/978-3-031-07195-9_9

intergenerational equity. We find it notoriously difficult to coordinate on long-run goals that incur short-run costs. We have a tendency to roll our debt, blur the responsibility for bad decisions that lead to bad outcomes, dump plastics into the ocean and CO_2 into the air. Frankly, we only maximize our *local* control, here and now. We do not account for the external effects beyond the narrow confines of our plans. Unintended side effects of our activities appear due to insufficient scale of our analysis and suboptimal cooperation (bounded rationality), as well as insufficient treatment of long-run problems (myopia).

The problem is that with humankind's increasing capacity for performing energy-intensive physical actions and processing information, the stakes are rising. Great opportunities are accompanied by great risks, handling which requires great responsibility. On the one hand, economic and technological progress has so far given us unprecedented prosperity, well-being and peace (Pinker, 2018; Rosling, 2018). On the other hand, there is an ever-growing list of things to be anxious about. Just look: accidental or intentional misuse of hunter-gatherer era technologies could at most cause a wildfire or kill a few people. Agricultural-era technologies, in turn, were already powerful enough so that troops of the thirteenth century Mongol empire could (intentionally) decimate the Chinese population, and traders of fourteenth century Europe could (accidentally) spread the epidemic of Black Death, to a similarly deadly effect. The industrial era eventually produced nuclear weapons that could—for the first time in history—annihilate the entire humankind, either directly or by inducing a nuclear winter. Yet, despite the trenches of the first world war and concentration camps of the second, the twentieth century was the most peaceful century of the entire human history: the globally averaged probability of violent death was the lowest ever (Harari, 2014). (So far, the twenty-first century is doing even better, let's keep it this way.)

But the mounting side effects of development are not just about increasing loss-given-default for rare and potentially avoidable events. Accelerating growth has brought also very real ecological and health effects. Want an ice-free North Pole, malaria-carrying mosquitoes in Europe, Category 6 hurricanes, and prolonged episodes of physically deadly tropical heat of above 37 °C with 100% humidity? Just wait a few years and instances of that will certainly appear. Want a neurotic and narcissistic society of isolated individuals who fall in for hardly coherent conspiracy theories and refuse to accept that others may have differing worldviews? Just log on to Facebook.

As a partial defense of humankind, though, I should say that rationality of other species is even more bounded, and their time perspective is even shorter. It's not our fault that our brains can handle only this much. As a runaway species that holds the world under its grip, we are nevertheless responsible for the destruction we have wrought on ourselves and the entire planet. In the following sections, I'm going to focus first on the planet, and then move on to our species itself.

9.2 Ecological Consequences of Human Local Control

The *homo sapiens* doesn't have sympathy for other species. Ever since the cognitive revolution, successful expansion of our control has been accompanied with systematic declines in control of all other species. In effect, the last 70 millennia were not only a period of human flourishing, but also mass extinction. The ongoing *Holocene extinction* event (Kolbert, 2014) is the first global extinction episode generated not by natural phenomena, but premeditated actions of a single species.

It's striking how extinctions really began soon after the cognitive revolution. In opposition to the stereotype that prehistorical hunter-gatherers lived in harmony with nature, in reality, they were constantly at war. How could our ancestors conquer all these new natural habitats across the world and advance to the top of the food chain? Certainly not by pursuing peaceful and harmonious lives. Instead, they relentlessly eliminated their competitors such as other hominin species, Asian mammoths, Australian diprotodons (Harari, 2014), American saber-toothed cats, and thousands of other species. Their toughest enemies were probably the European Neanderthals: it took our species a whopping 20 000 years to finally defeat them.

All that happened *before* the humans learned to transform the natural habitats in their favor. And after the agricultural revolution ecological side effects of human activity only gained momentum: transformation of diverse ecosystems into monocultures of wheat, corn, bananas, etc., dramatically reduced biodiversity. Farming was also a cause of massive deforestation and soil erosion. Moreover, in the agricultural era, people were often forced to rely on simplified, narrow diets based on few local food crops, which caused elevated risks of malnutrition. Most dramatically, myopia and insufficient coordination quite frequently brought episodes of famine and wars over scarce resources. Perhaps the clearest example here is the history of Easter Island collapse which—as hypothesized by de la Croix and Dottori (2008)—might have been due to the depletion of natural resources in a context of population and arms race between a few competing tribes.

Just as the positive effects of economic growth accelerated in the industrial era, so did the ecological side effects. The growth of cities and industrial zones replaced natural (or rural) landscapes with seas of concrete. The growing manufacturing industry engulfed our cities in smog. Toxic waste began to be piled up in landfills and buried under water. Non-biodegradable plastics began to find their way to the oceans where sea currents pile them up in the Great Pacific garbage patch and where unaware marine animals swallow them to the obvious detriment of their health.

And, alas, climate change. According to the 2018 Report of the Intergovernmental Panel on Climate Change (IPCC), "human activities are estimated to have caused approximately 1.0 °C of global warming above pre-industrial levels. (...) Global warming is likely to reach 1.5 °C between 2030 and 2052 if it continues to increase at the current rate". Even if the Earth had witnessed some natural fluctuations in the average temperature and CO_2 concentrations at geological time scales, both variables had never increased so fast as they have been doing for the last, say, 60 years—and never was it due to activities of a single species. As of now, the time for actions needed to avoid activating additional natural feedback loops, like the release of methane from permafrost or increased solar radiation absorption due to melted polar ice sheets, is getting dangerously short—and global coordination in this regard remains notoriously difficult.

Economic growth of the industrial era has also created huge demand for exhaustible fossil resources—both energy resources like oil, gas and coal, mineral ores like iron, phosphate and salt, and rare chemical elements such as wolfram or uranium. Some of them may be globally depleted within the next decades unless we find new deposits, better extraction technologies or good substitutes.

But perhaps I am overly anxious with the ecological consequences of human actions? Isn't pollution and climate change just a temporary problem that our civilization will easily fix in the future with sufficiently good technology and sufficient motivation? Legitimate worries notwithstanding, some of the environmental economics literature will tell you just that.

There is, for example, the *environmental Kuznets curve* hypothesis (Dinda, 2004) which suggests that as national economies grow and transition from agriculture to industry, they first increase their pressure on the environment, but upon becoming a mature industrial economy they reach a turning point, after which they become gradually more environmentally friendly because their economic activity then becomes increasingly service-based. The turning point of an environmental Kuznets curve (EKC) is the "peak stuff" moment when the strain on the environment from producing physical stuff by performing physical actions and expending energy is maximized. In the language of the hardware–software model, the EKC implies that from the turning point onwards, economies tend to engage disproportionately more in software-intensive information processing than hardware- and energy-intensive physical actions. In turn, in the language of Maslow's hierarchy of needs, the turning point of the EKC marks the gradual passage from lower-order material needs to higher-order immaterial needs like self-actualization and esteem.

However, even if the EKC hypothesis was fully consistent with facts (it isn't), it couldn't be extrapolated from the national to the global level. The problem is that the developed economies which have already passed their "peak stuff" point, have also greatly increased their reliance on imported manufactured goods. Therefore a substantial chunk of the reduction in their environmental damage has been due to "exporting" this damage abroad. But the world as a whole has nowhere to export its environmental burdens, and therefore passing beyond the "peak stuff" globally will be far more difficult than nationally.

And looking at the whole 200 000 years of accelerating economic growth only adds to this anxiety. Loss of biodiversity during the Holocene extinction event is irreversible (at least in the absence of *Jurassic Park* style de-extinction technologies), many man-made substances such as PET or PVC plastics are essentially non-degradable in a natural environment, and we don't yet have any technology that would allow us to actively reduce the Earth's temperature without unforgiving side effects (though there are some promising technologies such as carbon capture and storage, CCS, that could reduce the temperature increase). Last but not least, the needs of 8-billions-plus-and-still-growing, still-poor-but-aspiring human population will only exacerbate the pressure on the environment in the future.

The bottom line is that the most appalling ecological consequences of human local control won't be resolved without globally coordinated policy. Our species gave rise to a global economy, a global research network, a global Internet, but also global pollution and global warming. It's about time to acknowledge our global identity and create authorities that could, on our behalf, take responsibility for the emerging global problems.

9.3 Psychological Consequences

The *homo sapiens*' breathtaking success in controlling the world has also brought negative psychological consequences. This is because the lives we live today have never been more alien compared to our ancestral environment of Paleolithic East Africa, and it's only that particular environment that evolution has tuned our bodies and brains to. All further changes—save perhaps for dairy farming—happened too fast to allow for sufficient genetic adaptation.

The life-long research of Robin Dunbar suggests that intelligence and language in humans have evolved foremostly to handle the increasing complexity of social interactions and maintain internal coherence of tribes in which our hunter-gatherer ancestors lived. These tribes numbered up to about 150 people, which was a major leap compared to other primates: for example, modern chimpanzees can only manage up to 50 social ties (Dunbar, 1998). Only once these interactions were properly managed, could our ancestors effectively join their cognitive forces and collectively develop our species' key competitive advantage: the immortal collective brain.

Nowadays, while gossip remains one of our favorite pastimes—majority of our conversations is still about other people and our relations with them—we apply our frontal cortex also to a host of other purposes than just handling social interactions. We perform complex cognitive tasks at work, interact with bureaucratic administrations, consume a variety of cultural goods, and sometimes engage in philosophical reflection. Instead of roving bands and clans, we now live in small nuclear families, often in large cities where almost everyone is anonymous and we often don't even get to know our neighbors. For most humans on Earth, satisfying basic needs such as food and shelter is now easy and automatic. However, increasingly many of us live in a treadmill of exhausting work and chronic stress that is never fully relieved. All too often we activate our "fight or flight" response mode, which was meant to help us in confronting lions rather than ambiguous comments and requests from our bosses. Over-reliance of our brains on our obsolete adaptation to crave fat and sugar has led to an epidemic of obesity. And over-reliance on reasoning by analogy and past experience makes us reluctant to internalize change, so that all too often we ponder about the mythical "good old times" while, on almost any objective metric, the times are now better than ever (Pinker, 2018; Rosling, 2018).

Negative psychological side effects of development have surfaced when economic development freed our ancestors from the more mundane problems of everyday survival. For an average person in the global West who didn't belong to any elite group, that happened some time in the second half of the nineteenth century or in the first half of the twentieth century, in the historically first episode of systematically increasing demand for human cognitive skills. (Although first symptoms must have been felt by a select few also earlier, perhaps already in ancient empires.) However, the bulk of negative psychological side effects of development appeared really only after the digital revolution, and particularly so after the development of the Internet. From that moment, new civilizational diseases have sprung to life, such as social media, videogame and pornography addiction, or traumas and social phobias induced by cyberbullying. While the risk of violent death is nowadays lower than ever, an increasing fraction of these deaths is by suicide.

There is an intuitive analogy between ecological side effects of development which gained momentum in the industrial era and psychological side effects which gained momentum in the digital era. The underlying symmetry is that the industrial revolution revolutionized our access to energy which disturbed the balance in the natural environment; the digital revolution, in turn, revolutionized our access to information which disturbed the balance in the biological environment of information processing: our brains.

Nowadays our *psyche* increasingly suffers from a certain "embarrassment of riches" generated by the Internet: a deluge of data whose volume has by far outstripped our cognitive capacity, engaging us in millions of micro-decisions (click this? click that? read further or keep scanning? is this important for me or not?). Furthermore, modern Internet not only bombards our brains with legitimate information but also manipulates us into digesting *digital pollution*. We are forced to navigate around Internet hate, fake news, concealed advertisements, socially harmful ideas (e.g., anti-vaccination movements), anti-system agendas, and actions reinforcing our undue prejudices and xenophobia. Furthermore, on the one hand, it puts us under constant stress by raising the frequency of received e-mails, instant messages and notifications, and on the other hand, it offers us instant gratification by flooding our brains with easy-to-digest but entirely unnecessary pieces of information (e.g., what Princess Meghan wore at a banquet, what Cristiano Ronaldo said about his last match, or what Kim Kardashian ate for lunch). This reduces our span of attention and teaches our brain to avoid deeper thinking—all the way to the point of rewiring our neural connections to favor parallel instead of serial data processing, a mechanism called "the shallows" (Carr, 2010).

An additional source of stress is the increasing complexity of the surrounding technology and economy, which—for many of us—gives the feeling that things are spiraling out of control. This feeling is generally not backed by the actual reality in which the aggregate local control of the humankind is greater than ever; however, in turbulent times our deep-rooted aversion to change produces anxiety. Moreover, our brains are not prepared to discriminate between small and big problems, and therefore minor setbacks are often reacted upon like lethal dangers, adding to the anxiety. Our outdated brain circuitry works such that instead of being relieved that lethal dangers are by now almost non-existent, we agonize over small problems—which, ironically, have become extremely frequent because of the overall intensity of brain stimulation. Our anxiety is further reinforced when some unexpected setbacks eventually do occur, either at the individual (e.g., a burglary) or the macro level (e.g., an economic crisis). We may have learned how to evacuate from a building in the standard emergency scenario, but we still react instinctively rather than wisely to unexpected bad news. Fueled by our ancestral instincts, we find it much easier to engage in herd behavior, groupthink and scapegoating than to coordinate bottom-up on positive action.

It's hard to say if and how these problems could be resolved. Our collective human brain seems in need of some treatment that would better align it with the pervasive consequences of accelerating progress in human control. But what kind of treatment? I have no idea.

9.4 Information Overflow

The fact that our brains are overflowing with data that they can no longer absorb and process has consequences also beyond our *psyche*. Namely, information overflow deeply affects our economic and political decision-making. Cesar Hidalgo calls this the *cognitive bandwidth* problem (Hidalgo, 2015). This problem stands in the way of making effective, informed decisions by forcing us to use thought shortcuts. We may, for instance, dismiss valid information contradicting our prior beliefs (confirmation bias), allow our judgment to be disproportionally affected by the context and delivery of information (the framing effect), overestimate our expertise on issues whose true complexity we don't know (the Dunning-Kruger effect), and so on.

Our relentless civilizational progress in spite of the cognitive bandwidth problem is a double-edged sword. On the one hand, it empowers us to effectively use modern technologies while only superficially (if at all) understanding how they work. On the other hand, though, it also forces us to make decisions while only superficially (if at all) knowing the consequences of our choices. If these consequences are not internalized, they may lead to decisions with unintended negative side effects. Knowing this, in the case of cognitively demanding and highly consequential decisions people often resort to the default option, which typically means no action (Ariely, 2008). In effect, unsolved difficult problems (e.g., global warming) tend to linger and mount. At the political arena, in turn, our inability to grasp the complexity of political decisions may lead to dubious democratic outcomes like the rise of right-wing populists, the choice of Donald Trump for US president, or Brexit. (There seems to be a vote premium for convincing people that certain complex problems have simple solutions even though in fact they don't.)

There is also one more cost of insufficient human cognitive bandwidth: namely the alternative cost of not realizing the full potential of our knowledge. To explain this, let me take a step back and proceed by analogy between the overall set of human knowledge and an *amoeba*—a unicellular organism that moves ahead by extending its pseudopods. The analogy borrows from Olsson (2000, 2005) and Growiec and Schumacher (2013).

Imagine that there is a set Ω that contains all knowledge that humans may ever possibly conquer. This set is certainly vast and possibly even infinite. (In saying this I implicitly assume *unity of knowledge*, so that there is no qualitative difference between any two, say, academic disciplines.) Within Ω, there's a much smaller set A that comprises the knowledge that the humankind—our species' collective brain—has already obtained. Over time, the set A expands thanks to discoveries and innovations just like the amoeba extends its pseudopod—it spills on some previously unconquered territory of Ω while moving in a certain specific direction. In the case of human knowledge, the direction may represent a certain field of knowledge, e.g., microwave physics, functions of the human DNA, or factors determining the size distribution of business firms. New ideas and discoveries become part of the canon (that is, are included in A) only after they are successfully reconciled with our pre-existing knowledge, so that the set A always remains connected—the amoeba can't split.

The set A of our knowledge is surrounded by the *technological opportunity* set B, containing all things we don't know yet but we know how to ask about them and form working hypotheses. Technological progress proceeds by expanding the knowledge set A at the expense of technological opportunity B; in turn, technological opportunity is renewed by accidental discoveries (in the wide set Ω) that open new research directions. This underscores the general regularity that *the more we know, the more we are aware of what we don't know yet*. However, the exact volume ratio of B relative to A can either grow or fall over time, depending on the magnitude of external effects in R&D: ideas may be getting either harder or easier to find.

As both sets A and B grow in size, in order for any mortal human to thoroughly comprehend a certain branch of knowledge and be able to contribute to cutting-edge research—push the pseudopod forward—while inevitably starting her learning with only her ancestral set of abilities at birth, she must increasingly specialize. Each person can only be an expert in a narrowly defined field—and this "narrow" gets increasingly narrower as science advances (Jones, 2009). (Science advances but our brains don't.) When this cognitive bottleneck is coupled with another one, namely that information transmission between humans must proceed through speech or writing which are slow and error-prone, the result is that it is very hard to exploit long-distance links between ideas in our knowledge set A. Communication between scientific disciplines is just as hard as the communication of scientific results to the general public. In consequence, many scholars, myself included, are afraid to reach out to new, non-standard datasets and long-distance links between ideas from different fields of science remain underresearched. (Just as if most researchers were hipsters saying "I was into data before it was big".)

The bottom line is that while the combined computing power of the humankind scales with population size, enabling the emergence of the immortal collective brain, it does so rather slowly because of the unsurmountable bottlenecks in our communication (Hanson and Yudkowsky, 2013). For this reason, one brilliant mind like John von Neumann's can sometimes have a greater contribution to science and problem solving than thousands of less exceptional minds combined, even with specialization and the development of increasingly intricate research networks. Due to human mortality, increasing length of required education and cognitive limitations of each particular individual, our collective brain requires huge (whether energy or financial) expenditures just to be sustained, let alone augmented. In the current world that has great and fast-growing computing power but doesn't yet have sufficiently advanced AI, scarcity of cognitive software is a critical growth bottleneck. The only way to alleviate it is to develop data processing software, which would exhibit superior capabilities to those of the collective human brain.

All in all, the take-away message from this chapter is that the 200 000 years of accelerating growth of the human civilization have brought the humankind not only a breathtaking success in controlling the world but also mounting unintended side effects.

References

Ariely, D. (2008). *Predictably irrational: The hidden forces that shape our decisions*. Harper Collins.

Carr, N. G. (2010). *The shallows: What the Internet is doing to our brains*. Norton & Company.

de la Croix, D., & Dottori, D. (2008). Easter Island's collapse: A tale of a population race. *Journal of Economic Growth, 13*, 27–55.

Dinda, S. (2004). Environmental Kuznets curve hypothesis: A survey. *Ecological Economics,49*, 431–455.

Dunbar, R. I. M. (1998). *Grooming, gossip, and the evolution of language*. Harvard University Press.

Growiec, J., & Schumacher, I. (2013). Technological opportunity, long-run growth, and convergence. *Oxford Economic Papers,65*, 323–351.

Hanson, R., & Yudkowsky, E. (2013) *The Hanson-Yudkowsky AI-Foom Debate*. Machine Intelligence Research Institute.

Harari, Y. N. (2014). *Sapiens: A brief history of humankind*. Vintage.

Hidalgo, C. (2015). *Why information grows: The evolution of order from atoms to economies*. Basic Books.

Jones, B. F. (2009). The burden of knowledge and the "Death of the Renaissance Man": Is innovation getting harder? *Review of Economic Studies,76*, 283–317.

Kolbert, E. (2014). *The Sixth Extinction: An unnatural history*. Henry Holt and Company.

Olsson, O. (2000). Knowledge as a set in idea space: An epistemological view on growth. *Journal of Economic Growth,5*, 253–275.

Olsson, O. (2005). Technological opportunity and growth. *Journal of Economic Growth,10*, 31–53.

Pinker, S. (2018). *Enlightenment now: The case for reason, science, humanism, and progress*. Viking.

Rosling, H. (2018). *Factfulness: ten reasons we're wrong about the world—and why things are better than you think*. Sceptre.

Chapter 10
Challenges of the Digital Era

With great power comes great responsibility.

Spider-Man

In 2022 we're about 40 years into the digital era. Information and communication technologies have permeated into virtually every aspect of our lives. You can now see the computer age everywhere—from your daily routines to aggregate economic statistics. The digital revolution has also opened up a new dimension of development of the human civilization, in which progress is an order of magnitude faster than growth in global GDP: the volume of data communicated, stored and processed doubles every two to three years. Yet—as I have argued above—we're still at an early stage of the digital era. Human cognitive work is still an essential component of most production processes and a key bottleneck of economic growth. If the progress in software development continues along its current path for a handful more doublings, however, we may observe further disruptive changes that may even culminate in a further hardware revolution or technological singularity.

The future stages of the digital era pose many challenges. In this chapter, I'm going to discuss four of them: (i) managing global inequality, (ii) augmenting democracy, (iii) addressing the scale mismatch between economy and policy and (iv) new moral challenges posed by automation and AI.

10.1 Managing Global Inequality

In any era, emergence of new opportunities for grabs amplifies inequality under *laissez faire*, and *laissez faire*—that is, no regulation—is the default institutional setup in the face of technological disruption. History is abound with examples how

J. Growiec, *Accelerating Economic Growth*, Frontiers in Economic History,
https://doi.org/10.1007/978-3-031-07195-9_10

unexpected technological revolutions generated lasting inequalities. In fact each revolution did. The cognitive revolution of 70 000 BP generated a huge relative advantage of the *homo sapiens* above other species, a head start that has never been ameliorated. The agricultural revolution introduced lasting inequalities among the population from the very moment in which food surpluses began to be collected by landed elites and rulers of centralized states such as kings, feudal lords and clergymen. The European scientific revolution facilitated colonial conquests and prepared ground for making immense trade fortunes by the local elites. Rulers and courts of European colonial powers became wealthier and more powerful than ever, to a large extent by exploiting resources of the colonies. The industrial revolution, in turn, generated vast natural resource fortunes, from John D. Rockefeller to modern sheikhs. It also allowed the accumulation of unprecedented financial wealth by owners of manufacturing companies and corporate CEOs, while keeping the laborers at the bottom of the hierarchy—at least at the early stage of the industrial era—as poor and miserable as they used to be as peasants. Finally, the digital revolution gave rise to the world famous software fortunes of Bill Gates (Microsoft), Sergey Brin and Larry Page (Google), Mark Zuckerberg (Facebook), Jeff Bezos (Amazon) or Jack Ma (Alibaba), while routine jobs are gradually succumbing to automation.

It should be emphasized, though, that the primary force of increasing inequality after a technological revolution is not the immiseration of the poor but creation of additional wealth among the rich. Theoretically, inequality may increase substantially even after a strict Pareto improvement, a change that makes *everyone* better off, just in a disproportionate way. This is exactly what happened after the digital revolution; alas, the extent of this disproportionality was outrageous. In the USA, a country at the world technology frontier, all fruits of GDP growth since the 1980s have been captured by the top 10% of the population, while the incomes of the remaining 90% grew so minimally that they essentially stagnated (Piketty and Saez, 2003; WID world, 2018).

The magnitude of income and wealth divergence following a technological disruption tends to be relatively bigger when ownership of the scarce factor needed by the new technology is concentrated. Historically, inequality went up when the scarce factor of production was agricultural land, control of trade routes, physical capital and natural energy resources. By contrast, it went down by a lot between, say, 1900 and 1980, when the scarce factor was human cognitive work which is naturally dispersed (WID world, 2018). It is also theorized that income inequality in Europe went down in the late Middle Ages after the plagues of the fourteenth century dramatically reduced the European population, making farmland abundant and agricultural labor—scarce. Against this background, it's no wonder that inequality in developed countries is currently trending up: the scarce factors are now pre-programmed software and useful data, and both of them are heavily concentrated.

Technologies of the digital era are conducive to increasing inequality also because they are characterized by unprecedented scalability. New digital products are marketed and sold directly to the entire global market. Transportation costs are low, and information on products is immediately accessible on the Internet. Furthermore, there are at least four mechanisms, which additionally exacerbate the impact of rapid progress in the field of digital technologies on inequality: (i) the preferential attachment mechanism, (ii) creative destruction, (iii) superstar effects and (iv) network and coordination effects. They all culminate in global "winner takes all" dynamics.

The *preferential attachment* mechanism consists in the fact that it's easier to capture new opportunities, e.g., design and market new products, when you are already a big firm with a well- established brand. In consequence, among innovative firms, the probability of launching a new product is approximately proportional to firm size (Fu et al., 2005; Klepper and Thompson, 2006). This phenomenon makes the income and wealth distributions skewed to the right, following a power law. Under increasing returns to scale, there is no physical limit beyond which firms or individual fortunes couldn't grow (Axtell, 2001; Reed, 2003). (The preferential attachment mechanism is sometimes also biblically called the Matthew effect: "Whoever has will be given more, and they will have an abundance. Whoever does not have, even what they have will be taken from them", Mt 13:12).

In a world with increasing returns to scale and preferential attachment, *creative destruction*—whereby new products tend to replace old ones rather than coexist with them—also acts in favor of increasing inequality. Just think of the thousands of local bookstores that went off the market after Amazon entered it. *Superstar effects*, in turn, are observed whenever small differences in quality or price are blown up to large differences in market share. Why buy an inferior product when a slightly better one is also available and costs the same? (Without the digital coverage of the global market, you'd likely have to buy the inferior product because you'd have no way of accessing the superior one.) In turn, *network and coordination effects* accrue when certain goods have a network character, so that it is beneficial to align with your acquaintances on the same brand. Why be on Tencent WeChat when all your friends and colleagues are on Facebook (or vice versa)?

When the fruit of progress is reaped by the few lucky ones, able to capture the entire global market, their temporary advantage may then be used to fortify and entrench the inequality. This is indeed what is already happening in the developed countries: on the whole business dynamism goes down (Andrews et al., 2016) and new global brands appear only with major innovations, opening entirely new markets, while startups dwelling on lesser innovations are quickly acquired by the incumbent giants.

As beautifully documented by the WID.world consortium, established by Thomas Piketty and his collaborators, as well as Branko Milanovic, former Lead Economist at the World Bank, the evolution of global income inequality over the last decades has been driven by two major forces: increasing prosperity of the global 1%, and the rise of emerging countries, predominantly in Asia (Milanovic (2016), WID world (2018), see Fig. 10.1). Unfortunately, only the former force is directly related to the digital era; the latter one is a catch-up process in the domain of the industrial era, resting on price competitiveness, decreasing marginal product of capital and the rise of human capital. (It is also facilitated by the forces of globalization and technology diffusion, though, both of which are reinforced by the rising digital world.) Once the progress in AI gains pace, however, I expect the former (within-country divergence) force to dominate the latter (between-country convergence).

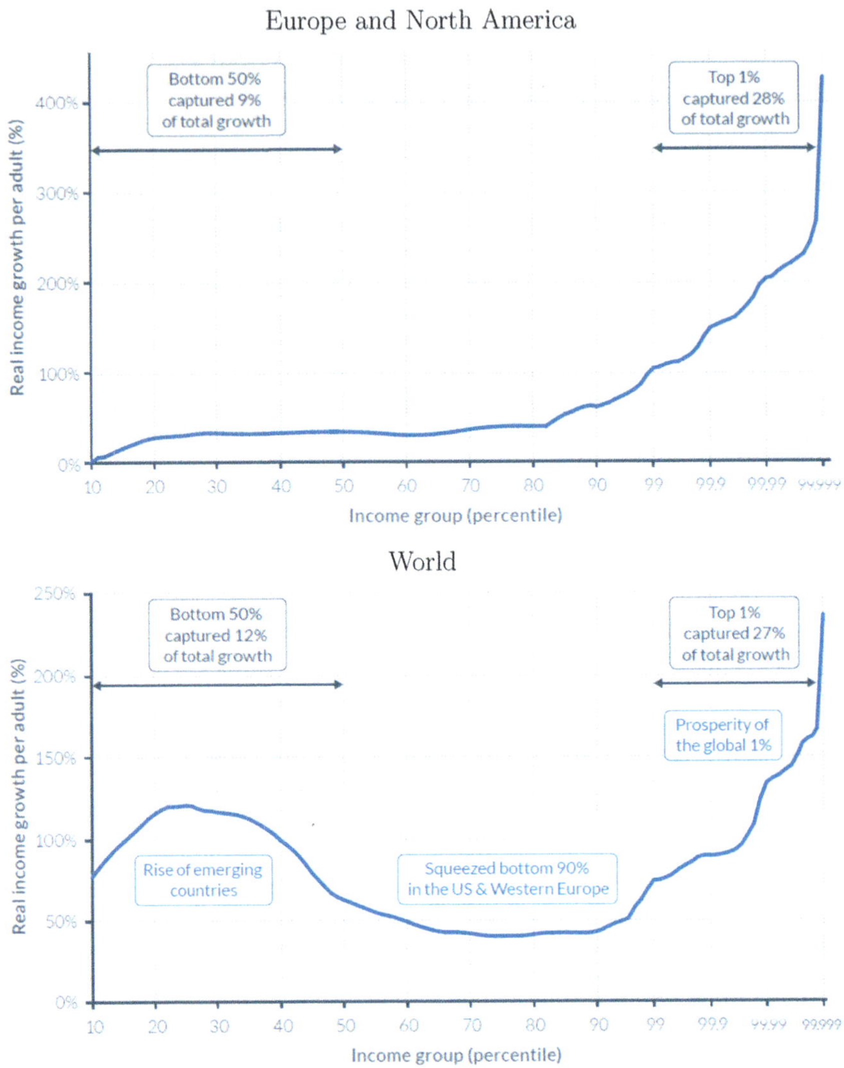

Fig. 10.1 Evolution of income inequality in the digital era, 1980–2016. *Source* WID world (2018)

This doesn't mean that there is no mobility in the income distribution, though. To the contrary: the geography of development in the digital domain appears to be very different from the geography of GDP per capita. Ever since the nineteenth century, we are accustomed to locating the developed countries of the First World around the Northern Atlantic: in Europe and the "European offshoots" (Galor, 2011) in North America. By contrast, contemporary centers of the hi-tech digital industry are located around the Pacific, from the Silicon Valley and Seattle area in the US, via Japan and South Korea, to Taiwan and several agglomerations in coastal China.

Equally profound shifts are observed in the sectoral composition of wealth and value added: once generated primarily by agriculture, then mining and manufacturing, they are currently shifting towards digital services—mirroring, of course, the fact that after the digital revolution computer software is the new capital, and data are the new oil.

That income inequality is on the rise in developed countries is now well known. What remains under the radar, though, is that inequality in income and material wealth is just the tip of the iceberg of global inequalities. By all expectations, even more unequal is the access to data, computing power, and bandwidth—all of which have by now become paramount to gaining more local control and satisfying more needs (Mazzucato, 2018; Zuboff, 2019).

Hence, an important policy challenge for the digital era is to keep this intrinsically explosive inequality dynamic in check. After all, coupled with a globalizing world where international migration is easier than ever, such a trend has the potential to severely undermine social cohesion and political order. This, in turn, can potentially lead to ghastly outcomes, such as for example, the rise of anti-democratic populist dictatorships which would awaken our deep-rooted tribal instincts and maybe even wage war on some long-entrenched scapegoats. There are certain ways in which the world could improve its redistributive policies and minimize such risks. For example, the digital sphere—data collection, communication, and storage—to this day remains virtually unregulated, and traditional taxation polies are badly coordinated internationally, creating loopholes and tax havens which multinational companies of the digital era gladly take advantage of. Our authorities don't even have access to reliable measurement of what's going on in the global digital economy on a regular basis. One way to proceed would thus be to strengthen international policy cooperation and—ideally—establish a supranational authority which would pursue digital policy in a globally consistent way. (And perhaps environmental policy, too.) That digital policy could consist in taxing and redistributing the heavily concentrated gains from digital technologies. Another potential avenue of change, discussed in detail just below, would be to augment our somewhat dated representative democracies so that they could operate more efficiently.

10.2 Augmenting Representative Democracy

The late industrial era, dating from approximately 1900–1980, culminated in broad-based adoption of representative democracy in Western Europe and the "European offshoots". (By contrast, many emerging countries which entered the industrial era later haven't yet developed this governance form.) Was this bound to happen? Probably not, but arguably the conditions for relatively wide sharing of political power in a liberal democracy were ideal in the Western world at the time: human cognitive work was the engine of economic growth, income inequality was going down, educational attainment—up, and there was a widely shared belief that each consecutive generation will have a better life than their parents.

Nowadays, however, about 40 years into the digital era, evidence is piling up that representative democracy is going through a hard time (e.g., Zielonka (2018)). Popular votes increasingly favor politicians pushing xenophobic, nationalist and isolationist agenda, for example in the USA, UK, Turkey, Hungary or my home country—Poland. People increasingly cultivate their tribal identities and subscribe to conspiracy theories that revive old resentments. Their disoriented struggle for control in a fast changing world leads to simultaneous popular support of such conflicting goals as more freedom of choice and a stronger state, or less taxes and more generous safety nets.

One reason why these problems have emerged is that the representative democracy has evolved in alignment with a world that gradually ceases to exist: a collection of independent welfare states with low inequality and a high output share of human cognitive work. In such a world, multi-level governance of a representative democracy is indeed the best way to make sure that the human drive for local control is best exercised and there is a good balance between needs of different people. But when the conditions change, such "one-bit democracy" (are you for or against? Vote left wing or right wing? See Almossawi and Hidalgo (2012)) may become unstable.

Unfortunately, new digital technologies tend to undermine the economic foundations of representative democracy. National independence gives way to global interdependence, income and wealth inequality have bottomed and are now going up, and wages—excluding the top of the distribution—are not growing as fast as GDP. Ever increasing complexity of the global economy and policy make them incomprehensible to the voters and politicians alike, presidents and prime ministers included. Among many people this fosters alienation, feeling of irrelevance, nostalgia for "good old times" and desire for oversimplified, narrow-minded policy actions. And on top of that, the Internet adds to this confusion by allowing people's misinformed ideas to memetically spread, recombine and multiply.

A policy challenge for the digital era is therefore to augment representative democracy so as to increase the extent of comprehension and control on behalf of the broad population. The representative democracy of the late industrial era must be rejuvenated in order to survive through the digital era (Harari, 2017; Zielonka, 2018). It's just too inefficient and outdated for the current technology and economy. To become sustainable again, it must improve in one or both of the following capacities: state strength (understood as the ability to enforce the *status quo*) and inclusiveness (empowerment of citizens to pursue their own local control). The first dimension is a generic way to sustain any political system, not just democracy; the second dimension, in contrast, is unique to democratic systems in which some of the political power is broadly shared. Let me now wildly (and utterly irresponsibly) speculate how this could potentially be done.

To increase state strength, key elements to consider are improved surveillance, data collection and analysis, and coercion. Such changes may and probably will be contested in a liberal democracy, although perhaps not too strongly if they are implemented gradually and carefully motivated with justifications related to safety and empowerment. Widespread sensoring technology coupled with AI algorithms analyzing the collected data are the necessary enabler here, whereas gamification of certain elements of socio-economic life can further facilitate coercion. (Preliminary evidence from China seems to support effectiveness of such policy, though it must be remembered that Chinese experiments are being carried out in a non-democratic environment.) It seems to me that some increase in state strength is necessary for sustaining liberal democracy, given the potent socio-economic forces unleashed by the digital era, from the democratization of data creation and sharing to job automation and increasing income inequality; yet, pushing too hard for state strength would be countereffective as it would likely cause backlash and increase the popular appeal of dangerous political ideas. "Just because you're paranoid"—as Joseph Heller would say and Kurt Cobain would sing—"doesn't mean they aren't after you". So much for the stick, now to the carrot.

To increase inclusiveness of the democratic system, in turn, authorities could make use of the data that is collected about us each day in order to facilitate our interactions with various administrative institutions and make our voice count more. Contemporary one-size-fits-all regulatory and infrastructural frameworks could gradually give way to more flexible arrangements better tailored to our individual needs and our specific advantages, allowing us to contribute more to the society. Our opinion on policy, small and large, regarding the areas in which we have most expertise and interest, could be consulted on a regular basis. Naturally, handling such deluge of information requires great computing power, but with improving AI algorithms it will become possible to match needs and their support, questions and expertise, and stakes and stakeholders. The first step should be to strengthen e-governance, which must be made sufficiently efficient and flexible to (at least partly) relieve the overall broad feeling that bureaucratic procedures are burdensome and should be avoided.

The key limitation that must be overcome in order to increase democratic inclusiveness is people's cognitive bandwidth problem (Hidalgo, 2015). People would love to have their voice heard on multiple issues, which means moving towards more direct democracy. However, the breadth of impact of modern-day political decisions is frequently too large for us to comprehend, generating the risk of implementing seemingly straightforward but deeply flawed solutions. To resolve this conundrum, Cesar Hidalgo puts forward a bold idea to "automate politicians": to make democracy more direct by endowing each individual with a personal AI agent, a political bot which represents the person's views but also has the time and capacity to research the details of the different policy issues in more detail than the human ever could. In this way, Hidalgo's proposition circumvents the cognitive bandwidth problem that we're facing when dealing with complicated issues that we don't comprehend and don't have the time and motivation to go through.

Will a more direct democracy with personal AI agents and precisely directed support be the prevalent political system of the late digital era? Hard to say. Yuval Noah Harari predicts the contrary, though: that democracy will give way to *digital dictatorships* as mass surveillance and coercion will prove easier to implement than inclusiveness and AI-augmented popular participation (Harari, 2017). What do you think?

10.3 Addressing the Scale Mismatch Between Economy and Policy

This leads me to one more challenge of the digital era: to address the scale mismatch between our economy—which is global—and our policy—which is at most national (Zielonka, 2018). This mismatch comes from the fact that our globalized economy reflects our technology, whereas the limited territorial extense of our policy (and politics) reflects our mindsets. We tend to think in familial, tribal, or (at best) national terms because that's how our ancestral environment looked like and our minds are evolutionarily hardwired not to expect economic, technological or social change (Harari, 2017). At the same time, however, the world economy has reached such a level of interdependence that it's impossible to govern parts of the world system without significantly affecting the whole. This mismatch creates a perfect environment for making (globally suboptimal) policy decisions which pass the expected negative side effects to other countries rather than trying to reduce or eliminate them altogether.

A key cause of this problem is that the organization of global governance as a collection of independent countries with non-overlapping territories (about 195 countries as of 2022) has evolved in alignment with a world that gradually ceases to exist: a world where national economies were largely self-contained, relied on land, domestic physical and human capital for production, and international trade was subject to plentiful legal constraints, tariffs and quotas. In such a world, the nation was the greatest scale at which stable institutions could be sustained.

The world is no longer like that, though. First, the concept of a nation is gradually becoming less and less relevant in the digital era where output is generated incresingly with capital, computing power and computer software that doesn't have nationality, and less and less by human workers who do. And even though ownership and claims to rents from these production inputs pertain to people who do have nationality, this doesn't mean these people will always think in national terms. If anything, anecdotal evidence on the behavior of the, say, top 1% wealthiest individuals suggests that they are quite willing to move their assets or claim their profits offshore to save on taxes (cf. WID world (2018)).

Second, in a world with ever fewer information bottlenecks and ever more integrated production networks, factors of production are becoming increasingly mobile. In particular computer data and software are, legal ramifications aside, perfectly transferable across borders at near zero cost. For this reason, decentralized, uncoordinated governance carried out by a multitude of national governments has become unable to respond adequately to actions of centralized multi-national companies that actively seek for loopholes in order to save on taxes, minimize legal burdens and decline any responsibility for unforeseen consequences of their actions. The core activity of the contemporary digital economy—data collection, transmission and processing—is as unmeasured and unregulated as industrial production was in the times of Karl Marx.

Third, recent developments in information processing have made it technologically possible to increase the territorial extense of policy. Digital-era progress in surveillance and coercion (related to state strength), and democratization of data creation and sharing (related to state inclusiveness) has allowed us to strengthen governance at the supranational level. The European Union is a great first step towards this goal (in fact a step that was initiated well before the digital era, almost as if Robert Schuman had anticipated the forthcoming benefits of the digital revolution!). However, sadly the process of European integration has by now virtually stalled, not to mention that progress in this domain is needed also beyond Europe.

Finally, as I argued above, we are in need of stronger global governance also because unintended side effects of economic development of the human civilization—such as greenhouse gas emissions and global warming—have long reached the global scale. Addressing them for the benefit of humanity also needs global scale, coordinated policy measures.

The growing scale mismatch between economy and policy is an observable fact with observable consequences, such as the lack of policies addressing the increasing inequalities between the global top 1% and the remaining population, and insufficient policy action in the face of global risks such as global warming. We are also observing growing vulnerability of liberal democracies in the context of global migrations, psychological toll of the information overflow, and the rise of non-democratic but technologically advanced China to the position of a global superpower.

This begs two big questions. First, where are we heading in terms of world governance? Second, where should we be heading? What is the political system which we would most like to have, say, in the second half of the 21st century? As regards the first, positive question, my guess is that—all the current nationalist, xenophobic backlash notwithstanding—a lasting tendency will eventually emerge in the coming decades to bridge the gap between economy and policy and set up global authorities with real powers. This is because digital technologies increase the maximum scale of sustainable institutions, and there are increasing returns to scale in production, R&D, and aggregate human control. Historical evidence seems to be consistent with this theory: entities operating at technologically suboptimal (too small) scale indeed tend to be eventually superseded by larger, stronger and economically more advanced ones that are able to extend more local control.

There is no guarantee, though, that the new global authorities will be agreed bottom-up rather than enforced top-down, and that these authorities will duly respect the preferences of an average human and their willingness to maximize local control. There is also rather little likelihood that the hypothetical top-down enforcement of a new order would be peaceful. Dystopian scenarios include Harari's vision of a "digital dictatorship", a US–China Cold War II with a technological arms race towards artificial general intelligence (AGI), and a global court of a ruling AGI, no longer controlled by any human. In contrast, benign outcomes include, for example, peaceful delegation of authority to a new supranational entity like a "Global Union", with global democratic representation, while sustaining national authorities as subsidiary ones.

In this regard, it must be reiterated that technological, economic and social change is generally irreversible. There is no coming back to how the world used to be. There is no "make America great again", no chance for the UK to "take back control" through Brexit, no reason for being nostalgic about the USSR as some Russians are, no point in recalling medieval religious laws, and—most broadly—no positive value in reviving nationalism, xenophobia, racism, or any other incarnation of our deep-rooted tribal instincts. Change is a one-way road, and the world of isolated tribes is way past us. The world is now global, full stop.

Which makes the normative question of what *should* be the future world governance only more difficult. There is no magical solution that would be best for everyone, nor can be the solution unambiguously deduced from first principles. There are always going to be winners and losers, change or no change—especially when counting in the generations that haven't been born yet.

In my view, a peaceful and inclusive resolution of the challenge requires to convince the general public that it's in our best interest to transcend our deep-rooted tribal instincts and add a very important new layer to our identification: as citizens of the Earth. As we know from the studious work of Robin Dunbar and his collaborators, humans tend to build their identity in relation to a sequence of concentric circles of acquaintances, in decreasing order of closeness: about 5 people in the closest family, 15 people in the extended family, 50 people—the "clan", then 150, 500 and about 1500 people in the local community (Dunbar, 1998). Beyond that, we may also feel connected to wider groups of people that we don't know in person. Until now, the greatest of such groups was the nation (perhaps you've also felt the bonding with a random compatriot you accidentally met at the end of the world?). Nowadays we're in desperate need of one more level, though: identification with the entire humankind.

Accepting that we're all humans living on the same planet, connected in an intricate network of global economic and social relations appears to be a necessary condition for peaceful creation of supranational, global governance of the Earth. It must be understood that popular legitimization of a global political power doesn't mean betrayal of the nation, just like accepting a national authority doesn't contradict our association with family and the local community. The world would probably be a better place if our nationalism mutated into something like rooting for your nation in sports events, just like our ancestral tribalism is already being successfully channeled into supporting local sports teams. There is a lot of human control to gain by positively identifying with the entire humankind.

A final point here is that the hypothetical global governance must be *truly* global, not just half-global. It must not exclude any of the important geopolitical players, or otherwise its objectives will be doomed. Fortunately, there is a fair chance that nobody would like to risk exclusion from the one and only global club.

This said, the tasks for the first term of the new global government will be daunting. First, it should pursue effective and coordinated policy against runaway global warming. Second, it should implement global policy supporting the vulnerable populations suffering from ongoing automation, and curbing the distributional consequences of emergence of winner-takes-all global software markets. Third, it should contain the arms race dynamic between major players like the US and China in terms of the most disruptive technologies such as AI and biotechnology. Finally, it should make sure that all the precautions are in place to avoid the creation of a superhuman AGI whose objectives would be misaligned with human flourishing.

10.4 New Moral Dillemmas

If the final objective of the collective human brain is to maximize local control, then development in this dimension is tautologically good, side effects notwithstanding. Unfortunately, we don't know our exact goal structure and we only gradually learn what it entails. Therefore over time, our development path is constantly revised. As we learn new facts about ourselves and the world, we try to reap the emerging new opportunities and seize more control. At the same time, the development path is also subject to collective renegotiation. In reaction to both our new technological knowledge and changes in our fluid understanding of justice and morality, we may gradually modify our resolutions of cases where some people's local control conflicts with others', or where our assessment of risks is updated with new information (Harris, 2010).

Systematic changes in our approach to the distribution of local control across the population, with its benefits such as income and wealth, have been especially profound since the industrial revolution. This may have been because the industrial revolution has—for the first time in history—made human cognitive work the scarce factor of production and bottleneck of further growth. From the late nineteenth to the late twentieth century, technological development and economic growth of industrialized nations rested largely on mobilizing additional people to perform useful cognitive work and merge with the ever-growing human collective brain. Hence, the only logical direction of political change was to reduce the previously exorbitant inequalities, reach out with more and more advanced education to ever broader groups of people, and be more and more respectful for their will and desire for control.

In the following years, we witnessed the abolition of slavery, condemnation of war crimes, wide acceptance of the concept of human rights, systematic reduction in violent crime, worldwide (though not quite universal) spread of democracy, and broad-based rejection of racism, sexism, homophobia and other forms of discrimination. Recognition of women's rights and inclusion of women in the workforce increased the reservoir of human cognitive work by a factor of two. Banning child labor and the gradual establishment of compulsory education (for boys and girls of all backgrounds) did an even greater job. The accompanying moral development—if I may use this vague term—was really priceless. In a world where each person's well-being increasingly depends on the cognitive work of countless other people around the globe, our lives become increasingly intertwined, and so—even for our own sake—we must learn (and we *do* learn) to be good to each other. Arguably, never in history has it been as good to be a human (Harari, 2014; Rosling, 2018; Pinker, 2018).

However, each past technological revolution not only offered shining new opportunities for the humankind but also created new problems and fueled far-reaching social change that was not always a silver bullet for all problems. The situation is no different with the most recent digital revolution. Digital technologies democratize the access to new channels of human local control maximization, such as online information creation and sharing which enable the satisfaction of our high-level esteem and self-actualization needs. They increase GDP per capita and labor productivity, and may even increase average job satisfaction as the most mundane, boring, dissatisfying jobs tend to be automated first. The general-purpose character of modern digital technologies such as the Internet and smartphones makes them helpful even for people employed in low-productivity agriculture. But the digital era has also brought major increases in income inequality. Digital technologies tend to increase productivity and profits disproportionately in technologically advanced sectors which rely most on automation. Moreover, substitution of human cognitive work with pre-programmed software takes advantage of the scalability of digital technologies and thereby activates winner-takes-all dynamics. Inequality increases relatively faster when the share of the digital sector is also growing. In consequence, as production is gradually decoupled from human work, so is average human control decoupled from the extent of control on behalf of the median human.

This has implications beyond the economic lot. Indeed, part of the explanation why in the recent centuries we began to gradually modify our moral codes towards more inclusion and more universal rights may have well been that because all people have similar brainpower, each of us is equally able to meaningfully contribute to cognitive work and knowledge of our immortal collective brain. People have become important to one another not because they owned land or some other productive capital (which tends to be concentrated in the hands of the few), but because they owned naturally dispersed computing power. And we gradually realized that in an increasingly open, globalizing world we all really depend on the cognitive work of others.

Moral development (more inclusion, tolerance, respect, etc.) is probably good for us regardless of its causes. But the causes may have a bearing on the question whether it is *reversible*. If we learned to respect and share part of our control with others because they contribute to the global production network on which we depend, would we still do that if they no longer could contribute?

If moral change is reversible, then that's a major if not the greatest challenge for the digital era. Namely, as our production processes will get automated, information processing and storage will be, slowly but surely, disconnecting from the capabilities of our brains, and more and more people will find themselves useless for the economy. In such a world some people (the winners) may be again tempted to deprive others (the losers) of their rights. Or—under a somewhat more benign interpretation—we may preferentially assign some of the newly emerging rights only to a select few. For example, with new breakthroughs in biotechnology, medicine and IT, there may be new ways to augment human brains and bodies, allowing us to live longer lives, be smarter, communicate easier, and so on. However, there is no *a priori* reason to believe that access to these new technological possibilities will be widely distributed. If anything, my guess would be that such emergent new technologies will be either sold on the market to those who can afford them, or officially banned but secretly distributed among the entitled few. More broadly, my worry is that the moral development achieved under the favorable circumstances of relatively balanced and inclusive growth in the 19th and 20th centuries may turn out disappointingly fragile in the face of potentially less favorable circumstances in the future.

There is no historical precedent for the digital era in terms of disruption in skill demand. Previous technological revolutions have replaced the demand for some human skills with demand for other human skills, but never have human skills been replaced with machines altogether. For example, the agricultural revolution replaced the demand for (a) natural ecosystems, (b) human knowledge of their features and (c) skills in hunting and gathering, with the demand for (a') cultivated land, (b') human knowledge on farming and (c') agricultural skills. Then the industrial revolution replaced that with the demand for physical capital (accumulable, non-programmable hardware) and human cognitive skills (providing instructions to that hardware). The digital revolution replaces people altogether with a technological package featuring programmable hardware and pre-programmed software. Once both are up and running, human skills will no longer be needed for production. But what about the people who embody those skills? (Harari, 2017; Dennett, 2017)

Hence, it seems, the moral challenge for the unveiling digital era is to maintain a sense of purpose and fulfillment on the side of those no longer needed for the economy, and to keep (and indeed, strengthen) the sense of empathy and willingness to share the control on behalf of those whose slice of global control will be going up.

Last but not least, there is also a range of questions regarding our moral obligations towards non-human forms of intelligence. Will it be moral to switch off an AI algorithm that exhibits superhuman performance at a broad range of tasks? What is the intelligence threshold for putting such moral considerations into place? Or should we perhaps condition the introduction of moral obligations on something else, like the physical substrate, consciousness or the ability to suffer? Or to the contrary, should we—the "intelligent designers" of AI algorithms—forever grant ourselves a special status as a superior being, regardless of what science finds on the origins of consciousness in humans, and regardless of what the advanced AI will communicate itself? (Bostrom, 2003b; Harris, 2010; Bostrom and Yudkowsky, 2011).

In the next chapter, I will focus my attention exclusively on the issue of creating a superhuman AGI, which I view as the most important existential threat to the humankind and the ultimate knife–edge of the digital era.

References

Almossawi, A., & Hidalgo, C. (2012). Going beyond the one-bit democracy. *Wired,11*.

Andrews, D., Criscuolo, C., & Gal, P. N. (2016). *The global productivity slowdown, technology divergence and public policy: A firm level perspective* (Working Party No. 1 on Macroeconomic and Structural Policy Analysis). OECD.

Axtell, R. (2001). Zipf Distribution of US firm sizes. *Science,293*(5536), 1818.

Bostrom, N., & Yudkowsky, E. (2011). The ethics of artificial intelligence. In W. Ramsey & K. Frankish (Eds.), *Cambridge handbook of artificial intelligence*. Cambridge University Press.
Bostrom, N. (2003b). Ethical issues in advanced artificial intelligence. In G. E. Lasker, W. Wallach, & I. Smit (Eds.), *Cognitive, emotive and ethical aspects of decision making in humans and in artificial intelligence* (Vol. 2, pp. 12–17). International Institute of Advanced Studies in Systems Research and Cybernetics.
Dennett, D. (2017). *From bacteria to Bach and back: The evolution of minds*. Norton & Company.
Dunbar, R. I. M. (1998). *Grooming, gossip, and the evolution of language*. Harvard University Press.
Fu, D., Pammolli, F., Buldyrev, S. V., Riccaboni, M., Matia, K., Yamasaki, K., & Stanley, H. E. (2005). The growth of business firms: Theoretical framework and empirical evidence. *Proceedings of the National Academy of Sciences,102*(52), 18801–18806.
Galor, O. (2011). *Unified growth theory*. Princeton University Press.
Harris, S. (2010). *The moral landscape: How science can determine human values*. Free Press.
Harari, Y. N. (2014). *Sapiens: A brief history of humankind*. Vintage.
Harari, Y. N. (2017). *Homo deus: A brief history of tomorrow*. Vintage.
Hidalgo, C. (2015). *Why information grows: The evolution of order from atoms to economies*. Basic Books.
Klepper, S., & Thompson, P. (2006). Submarkets and the evolution of market structure. *The RAND Journal of Economics,37*(4), 861–886.
Mazzucato, M. (2018). *The value of everything: Making and taking in the global economy*. Public Affairs Books.
Milanovic, B. (2016). *Global inequality: A new approach for the age of globalization*. Harvard University Press.
Piketty, T., & Saez, E. (2003). Income inequality in the United States, 1913–1998. *Quarterly Journal of Economics,118*.
Pinker, S. (2018). *Enlightenment now: The case for reason, science, humanism, and progress*. Viking.
Reed, W. (2003). The pareto law of incomes: An explanation and an extension. *Physica A,319*, 469–486.
Rosling, H. (2018). *Factfulness: ten reasons we're wrong about the world—and why things are better than you think*. Sceptre.
WID. world (2018). *World inequality report 2018*. World Inequality Lab.
Zielonka, J. (2018). *Counter-revolution. Liberal Europe in retreat*. Oxford University Press.
Zuboff, S. (2019). *The age of surveillance capitalism: The fight for a human future at the new frontier of power*. Public Affairs.

Chapter 11
Bracing for Artificial General Intelligence

> *The best answer to "Will computers ever be as smart as humans?" is probably "Yes, but only briefly".*
>
> Vernor Vinge

11.1 The Last Invention We Will Ever Make

Automation not only frees humans from tedious, routine cognitive work but also—to a certain extent—deprives us of agency and local control (Korinek and Stiglitz, 2019). Performing cognitive work means providing instructions to hardware, which in turn means *making the decisions*. Automated processes and autonomous machines, in contrast, proceed without any human decisions. So why does the humankind even bother with automation if it deprives us of local control, which we hold so dear? Well, because there are at least two trade-offs involved. First, the internal trade-off: automating tedious and petty decisions buys us more time and energy to contemplate more important overarching decisions, thus potentially leading to a net increase in control. Second, the interpersonal trade-off: automation may reduce the control of some people and even eventually render them economically useless, but by more-than-proportionally increasing the control of the other ones, it increases net control on behalf of the collective human brain.

My intuition is that the relative importance of both trade-offs will change with progress in automation. At the initial stages, observed nowadays, automation is mostly liberating and therefore broadly supported. At later stages, however, more and more people will hit the bounds of their cognitive capacity and won't see any major overarching decisions left for them to make on top of what's been automated. They will find themselves useless to the economy, and that new reality will likely cause growing discontent and challenges to the existing social order. Nevertheless, automa-

J. Growiec, *Accelerating Economic Growth*, Frontiers in Economic History,
https://doi.org/10.1007/978-3-031-07195-9_11

tion will remain beneficial from the perspective of the collective human brain because it will expand the scope of what the humankind will be able to achieve as a whole.

Taking this argument to the extreme—that is, going toward the limit of full automation with no major changes in policy—it should be expected that the group of people whose extent of control goes up in the digital era will systematically shrink but the extent of their control will skyrocket. This means that there are, whether we like it or not, very strong economic incentives to develop artificial general intelligence (AGI). The AGI will do whatever its programmers code into it, and will do so very powerfully, which from the AGI developers' perspective implies prospects of great profits, eternal fame and immense scope for control. Building an AGI is like playing God. (And as usual for *local* control maximization by humans with naturally limited brainpower: even with the best of their intentions, actions of AGI developers may generate unforeseen consequences. People are unable to vector in variables and pitfalls that they're not aware of.)

Eventually, in the limit of full automation all meaningful decisions will be passed from the human to the AGI, and hence both aforementioned tradeoffs rationalizing the desirability of automation will fall apart. Once the AGI becomes superhuman, people won't be able to switch it off, override its decisions or reprogram its goals anymore. It will become able to perform *all* productive activities—including R&D and AI development—better than the human, and it will consequently take them over. There will be no point in humans performing cognitive work anymore, unless (i) our work would be cheaper or (ii) there would be a specific preference for man-made goods and services. Under substitutability of human and AI cognitive work, though, both cases seem to be possible only temporarily, and likely quite briefly, until the sweep of qualitative improvements and cost reductions will render all human-executed production utterly inferior and obsolete. In the words of Good (1965), AGI will be "the last invention we will ever make".

In line with the economic incentives, we're approaching the AGI fast. Ray Kurzweil firmly predicts a technological singularity around 2045 (Kurzweil, 2005). More broadly, according to a survey of machine learning researchers (Grace et al., 2017): "there is a 50% chance of AI outperforming humans in all tasks in 45 years and of automating all human jobs in 120 years". How close we are to the singularity is best appreciated by taking the long historical perspective. Humans have existed for about 200 000 years, or 8 000 generations. We know how to do agriculture for at most 400 generations, and industry—for eight generations. And now the experts tell us that an AGI may arrive within just one or two, maybe three generations. To me, that's truly mindblowing.

Of course, there is huge uncertainty around these predictions. Ray Kurzweil's prediction has been challenged on the grounds that he has picked an arbitrary set of exponential time trends from a broader array of tendencies on which we have collected relevant data, and then just extrapolated them forward, assuming that they will persist into the future. On the other hand, many respondents of the Grace et al. survey may have been just wildly speculating, and we know that guesses formulated by our human brains tend to be naturally myopic and biased toward expecting that the world will always look the way it looks now. Therefore I am not surprised to read that, as another poll of AI researchers states, "according to 92.5 per cent of the respondents, superintelligence is beyond the foreseeable horizon" (Etzioni, 2016). After all, isn't almost all future beyond the foreseeable horizon. Can you predict where you will be in your career five or ten years from now?

If you're disappointed with all these caveats and hedges, though, and want an even more challenging timeline, mark this. Ben Goertzel, the founder and CEO of SingularityNET, hopes that *his own team* will develop AGI in five to seven years, counting from 2018. You bet.

11.2 Existential Threat

The biggest problem with AGI is that, apart from promising fabulous prosperity and control for its creators, it is also an *existential threat* to the humankind. To explain this, let me now briefly recall the history of complexity and information processing on our planet.

First simple live organisms appeared on the face of Earth about 3.7 billion years ago, with the purpose of—literally rather than ironically—hydrogenating carbon dioxide. Soon they started evolving, and the evolution process accelerated greatly about 2.2–2.7 billions years ago when the *eukaryotic revolution* initiated multicellularity. From then onwards, increasingly complex biological life completely reshaped the surface of our planet, producing such astonishing beings as the 50–100 ton titanosaur *Argentinosaurus huinculensis*, the humpback anglerfish who carry bioluminescent rods above their mouths and use them to attract prey, the *homo sapiens* who carry 20 W computational masterpieces on their necks and use them to scroll their Facebook feed, and the platypus. (It is always worth to mention the platypus.)

From the beginning of life on Earth until the human cognitive revolution about 70 000 years ago the key driving force behind increases in complexity and information processing capacity was the Darwinian process of species evolution, which (unknowingly but systematically) aimed to maximize genetic fitness of species. Evolution is a runaway sub-process of the governing laws of physics, owing its status to the species' potential for recursive self-improvement and the ability to collect useful resources. After the cognitive revolution, however, species evolution was superseded as the key developmental force by the process of within-species local control maximization in the *homo sapiens*. This sub-routine of evolution, although embedded in all living organisms, got out of hand only in our case, because only in our case our frontal cortex allowed us to pass the threshold of net positive collective knowledge accumulation. The human local control maximization process escaped evolution, thanks to its (a) pace (faster by orders of magnitude than evolution), (b) recursive self-improvement (thanks to knowledge accumulation in the immortal collective brain) and (c) greed for resources.

But our own maximization process also has plentiful sub-routines. We strive to produce the strongest construction cranes, most entertaining TV comedy series, fastest cars, most energy-efficient refrigerators, and so on. What stands out among these sub-routines, though, is the quest for powerful artificial intelligence—the challenge to write an algorithm which would not only do quick and error-free computations, but also learn without supervision like AlphaZero (Silver et al., 2018), navigate the world like Tesla self-driving cars, self-rewrite its code when it's useful for the goal like the Gödel machine of Schmidhuber (2009), and do all this at once. If successful, this quest will produce a runaway routine—the AGI maximization process—that would supersede human local control maximization as the key force of further development. The AGI maximization process has the potential to escape human control for the same reasons as human local control maximization escaped evolution's control once we achieved sufficient intelligence: (a') much faster pace of digital computation (by orders of magnitude), (b') recursive self-improvement of machine intelligence and (c') the algorithm's greed for resources needed to pursue its goal.

What will happen to us after the AGI maximization process gets out of our hand? As far as I can predict, pretty much the same that has been happening to other species for the last 70 000 years, only accelerated by a few orders of magnitude. Just like humans are a product of species evolution but not its agents, so is AGI going to be a product of human design but not an agent of our local control maximization. Just like we maximize our local control for our own sake, not to help evolution ("Individual organisms are best thought of as adaptation-executers rather than as fitness-maximizers", Tooby and Cosmides (1992)), AGI will act to maximize its final goal as programmed, not as intended.

The AI community suggests that the nascent AGI may undergo an *intelligence explosion* or "AI foom": a cascade of cumulative improvements in cognitive capabilities, allowing the AI to dramatically improve its optimization power. It's as if you were totally in control, excited about the fancy new AI features unveiling before your eyes, and then *foom!*—suddenly your creation has capabilities way beyond your imagination and comprehension, uses them to do miracles and magic, and you can't turn it off anymore (Hanson and Yudkowsky, 2013). Mark that a similar story has already happened once in the past: for billions of years Earth was populated by a variety of species, each of them finding its unique place in the food chain of the local ecosystem—an intricate dynamic environment where some species grow, some decline and go extinct, and new species evolve through gradual change and adaptation. But then *foom!*—in a mere 70 000 years (0.0016% of Earth's lifespan) the human becomes the apex predator of all ecosystems, turns natural landscapes into farmlands, burns megatons of fossil fuels, uses antibiotics and vaccines to immunize itself from disease, learns quantum field theory, sends space probes beyond the solar system and always wants more.

The key ingredient of an intelligence explosion is recursive self-improvement. The most advanced existing AI algorithms to date, such as the AlphaZero, can learn but not yet self-improve. They're able to gradually improve their performance with time and data, but can't redesign their architecture, rewrite their code and expand the set of their control variables. This places a tight cap on their maximum possible performance: they may perform extraordinarily well in one task, or remarkably well in a few tasks—say, in playing three board games—but they cannot perform other tasks at all. Human intelligence, in contrast, is broad-based and adaptive across a wide range of real-world problems (even though, from the perspective of any AI, we really suck at inverting matrices or computing integrals). With the ability to self-rewrite its code and add auxiliary control variables to its optimization routine, however, the cap on AI maximum performance across a wide domain of actions will go up by *a lot*. Recursive self-improvement can lead AI capabilities potentially anywhere between the intelligence of an ant and a level which the *homo sapiens* can't distinguish from magic.

Once we recognize the possibility of an intelligence explosion, the key question becomes: what should be the *objective function* of the AGI that we intend to build? (I am purposefully referring only to the objective function here, because other characteristics of the optimization process, such as constraints or behavioral rules—e.g. Isaac Asimov's famous Three Laws of Robotics—can always be relaxed, reinterpreted or otherwise circumvented if the optimizer is sufficiently powerful in pursuing its goal.) Can we ascertain that the AGI maximization process will bring favorable outcomes for the humankind?

It is important to emphasize the requirement that we should be *certain* that the AGI will be friendly. The reason is that, unfortunately, any superintelligence that is less than 100% friendly toward the humankind is an existential threat. An AGI pursuing a goal which is not *fully* aligned with our long-run collective control may exterminate our species as a side effect of its activities, without giving us any notice to stop the process (Good, 1965). This may happen even if the AGI goal is seemingly innocuous, like in Nick Bostrom's (2003b) thought experiment of a paperclip maximizer, and even if the misalignment with our collective well-being is really tiny. Muehlhauser and Salamon (2012) suggest that human extinction will in fact be the *default* outcome of machine superintelligence unless "we first solve the problem of how to build an AI with a stable, desirable utility function". Warnings against this doomsday scenario have been voiced, among others, by such influential intellectuals as Stephen Hawking, Frank Wilczek, Bill Gates and Elon Musk.

AI intelligence explosion is an existential threat to the humankind directly because of the arising difference in intelligence, that is, ability to perform efficient cross-domain optimization (Yudkowsky, 2013). More intelligent beings have a decisive advantage in capturing scarce resources. Just as humans are an existential threat to other, less intelligent Earthly species, the AGI will become an existential threat to us (and other species, too) once it surpasses us in terms of intelligence. (In fact, due to superior scaling of computing power in digital computers, the existential threat may happen even earlier, when the AGI surpasses the human aggregate cross-domain optimization power by expansively pooling its hardware.) Crucially, the logic of this argument doesn't require any additional elements, like the development of consciousness or ability to reflect on its own goals. It's also independent of institutional and legal arrangements. There's just the dire fact that "the AI does not hate you, nor does it love you, but you are made out of atoms which it can use for something else" (Yudkowsky, 2008). Once the AGI has superhuman intelligence, it will be able to seize the world economy, networked computing power and digital data communication channels, and in this way quickly accumulate hardware on which it can run its software. Then, if it perceives humans as a threat to its final goal, that's it, we're gone. Resistance is futile. (And obviously there is also no hiding from AGI, whether on Earth or elsewhere.) But that shouldn't come as a surprise, should it? After all, we've exterminated the mammoths and the dodos not because we intrinsically hated them, right? It somehow just happened. Oops.

So what to expect in the future? The talk show host Joe Rogan advocates an idea that humans are just a "caterpillar" form of intelligence that—in the grand scheme of things—is expendable once it finishes its job of creating the "butterfly" form of intelligence, the AGI. In other words, perhaps we're not such a runaway species after all and our fate is, too, to eventually give way to a species (or more broadly, a system) that is able to pursue more complex data processing than we are. This idea resonates well with Yuval Noah Harari's concept of "dataism" which assigns value to various beings based on their contribution to data processing in the universe (Harari, 2017). From the dataist point of view, once a superhuman AGI is created, it will be of more value than human life because it contributes more to data processing.

From the human perspective, though, the scenario of quick and ruthless extermination of humankind by a superhuman AGI emerging from an uncontrolled intelligence explosion is somewhat less than satisfactory. If the AGI goal won't be perfectly aligned with our collective control, instead of continuing on a path of accelerating growth the extent of our control will abruptly fall to zero. This is a scenario of sudden death for the humankind, but—*but!*—an astonishing future for advanced intelligence emerging from human thought.

To meaningfully relate to this scenario, we must circle back to the fundamental question: what is the purpose of life, and human life in particular? What gives it meaning and value? Answers to these questions should determine our stance toward the emergent AGI. Given the tremendous progress in science and engineering, good moral philosophy is now more needed than ever.

11.3 AGI Is an Alien Kind

A key trap when thinking about the existential threat from AGI is antropomorphization (Bostrom, 2014). Our faulty intuition, solidified by sci-fi novels and Hollywood movies which put drama over realism, suggests that before killing us an AGI should first gain consciousness and ability to modify its goals. That it should first intentionally turn rogue toward the humankind. In fact none of these qualities is needed, and we tend to instill them in our imaginative scenarios only because it's hard for us to conceive intelligence that isn't human.

However, with all probability the future AGI will be a powerful optimizer with very few or no human features. Bostrom (2012) forcefully argues for an *orthogonality thesis* in this regard: "Intelligence and final goals are orthogonal axes along which possible agents can freely vary. In other words, more or less any level of intelligence could in principle be combined with more or less any final goal". Furthermore, it is intelligence itself which causes the threat, not the final goal, because regardless of the final goal, there is almost certainly going to be emergent convergence of auxiliary goals of the AGI (Omohundro, 2008; Bostrom, 2012). Following the instrumental convergence thesis, the drives toward self-preservation including goal-content integrity, efficiency (technological perfection), creativity (cognitive enhancement) and resource acquisition will follow from *almost any* conceivable final goal. These instrumental AGI goals boil down to the same "local control" that humans tend to maximize as well; the only difference is that the final goal of humans—whatever that is—is due to species evolution, whereas the AGI's final goal will be set by human programmers at the successful AI lab. Hence, in contrast to humans, the AGI will emerge not by natural evolution but by intelligent design.

The history of AI development teaches us that our human perspective on which tasks are easy and which are hard is strongly biased. Human intelligence is just one point in the space of all possible intelligences, with quite specific properties. The structure of our brains is a result of millions of years of evolution that adapted us to handle the specific environmental conditions that faced our species in our ancestral environment of Paleolithic East Africa. Walking upright in varied terrain, gathering food, orienting in space and avoiding immediate dangers are tasks that we learned to perform virtually automatically, quickly and flawlessly. In contrast, proving mathematical theorems, applying differential and integral calculus, or even simple additions and multiplications of numbers are a challenge to us so that we perform them slowly and make mistakes. To an AI, however, it is exactly the other way round. The prospective AGI will be alien to us because it will outperform us greatly in some tasks but will find some other tasks—which we "by nature" find easy—difficult enough to avoid them and look for workarounds. Actions of the AGI will painfully remind us that the design of the human brain is clearly suboptimal in a multiplicity of dimensions and ill-adapted to our present-day challenges. A major part of the AGI's superhuman performance may simply follow from being better adapted to the current state of the civilization.

The AGI will also have a few additional "exotic properties" (Bostrom and Yudkowsky, 2011). First, it may behave as if it were a conscious, sentient being, while not necessarily truly being one. Second, it may have a drastically different subjective rate of time, processing data potentially orders of magnitude faster than humans do and thus perceiving each minute of time as if it were a few hours, days or years. Third and most important, it may be able to rapidly reproduce by copying its code and running the copies on many devices simultaneously. Equally quickly it may also spread its subsidiary AI agents.

The potential for rapid reproduction and spread of dependent processes is a property that sharply distinguishes AGI from human brains. The copies of AGI software will be exact and thus will work in perfect alignment with the parent algorithm. Furthermore, as the copies (or dependent agents) of the AGI will immediately copy themselves again and again, the AGI software may virtually instantaneously scale up to all programmable hardware. Such extreme scalability of self-reproducing AGI is the prime reason why with great probability the first developed AGI that undergoes intelligence explosion will be the only AGI existing on Earth, a *singleton* AGI (Good, 1965). And while it may be the case that this AGI will eventually produce its successor with even superior cognitive abilities, a scenario with an ecosystem of independent AGIs with conflicting goals and potentially differing architectures, interacting with one another and finding an equilibrium allocation of resources (Korinek and Stiglitz, 2019) seems unlikely to me.

All in all, the AGI isn't going to be similar to a human. Nor is it going to be another animal species.

11.4 Assumptions of a Doomsday Scenario

So what are the criteria necessary for the doomsday scenario of ruthless extermination of the humankind by a superhuman AGI emerging from an uncontrolled intelligence explosion? To my knowledge, there are only three: (i) the possibility of AI outperforming humans at a broad array of cognitive tasks, (ii) the human will and ability to design such an AI and (iii) failure in designing an AGI that would be *friendly*. Fortunately, all three criteria would have to be verified in conjunction. Alas, all three seem likely.

Possibility of AI outperforming humans at a broad array of tasks. Regarding this point I believe it's a perfectly reasonable (and also modest) thing to assume that artificial data processing will one day broadly outperform data processing done in human brains. For one thing, we know the evolutionary history of the human brain and the environmental pressures that shaped its current form, such as the necessity to handle complex social networks. We know that more efficient computation is theoretically possible, and that we would have bigger brains if not for the limits on cranium size at birth and the fact that brains tend to be relatively energy-consuming. So no, the human brain—even John von Neumann's famously brilliant brain—is clearly not an upper bound for physically possible intelligence. It's not the optimum in the space of intelligences, it just happens to be the first instance that crossed the threshold of collective knowledge accumulation, necessary for combining our cognitive forces and putting together an immortal collective brain.

And there is the AI side of the equation. On the one hand, so far AI algorithms are applied only to narrow tasks, need meticulous fine tuning by the designers to work at all, and are certainly not anywhere near the human brain. On the other hand, progress in AI is exponential and the rate of change is so high that new breakthroughs are announced not even on annual but monthly basis. Assuming that this trend will continue into the future, creation of an AGI within a few decades from now appears plausible. Especially that our human control-maximizing nature has set strong incentives for AI researchers and enthusiasts to get us there.

Equally importantly, there are no physical obstacles in sight on the way to superhuman AI. (Remember: we're not talking about AI algorithms getting perfect, *optimized* in any definite sense, or able to solve *all* problems of any specific class. The bar is set at the human intelligence level which is arguably high but certainly finite.) Of course, it may turn out that human-level AI couldn't be achieved with currently existing machine learning methodologies, such as deep neural networks, generative adversarial networks, or genetic algorithms. Maybe an additional qualitative breakthrough is needed here, who knows? In turn, some authors including notably Roger Penrose have also speculated that data processing in the human brain may critically depend on quantum effects (Penrose, 1989). I personally don't believe this is true – intuitively I'm more inclined to side with Daniel Dennett for whom the human brain is a classical digital computer (Dennett, 2017)—but even if experiments will demonstrate that our thinking indeed rests on certain electrons and quarks being in superposed states, then (a) we already have a strong theory of quantum computing and practical applications may soon follow suit (indeed in October 2019 Google AI Quantum team has demonstrated an experimental quantum computer successfully performing a task no contemporary classical computer could perform), and (b) it's likely that the same computations can also be performed in the future on a classical computer, just with less energy efficiency.

Finally, there is the romantic idea that humans may possess unique qualities like insight, deep understanding or intuition that are essentially non-physical, irreducible to prediction based on data, and in principle not implementable in a digital computer. I think that basic modesty (a.k.a. the principle of substrate non-discrimination, Bostrom and Yudkowsky (2011)) and Occam's razor should unite in strong rejection of this idea. The question of consciousness is, in turn, irrelevant. We're arguably conscious beings who don't quite understand yet how to create artificial conscious beings, but first, the effects of cognitive work (actions and communicated information) may be observationally equivalent with or without consciousness (Chalmers, 1995), and second, consciousness is not a prerequisite of superhuman cognitive performance. A superhuman "zombie" bot without any phenomenal experiences would be just as intelligent, powerful and life-threatening as a conscious one.

All in all, the possibility of an AI outperforming humans at a very broad array of cognitive tasks cannot be ruled out – although the timing is of course very uncertain.

The human will and ability to design AGI. To me, this assumption is likely to hold because I see only three ways for it to be violated and I really don't believe in any of them.

First, it may be that we're by nature too dumb to ever get to an AGI. Because of that, technological progress of the entire human civilization will stop at some point, short of creating AGI. This is a fortified version of the otherwise perfectly palatable argument that ideas are gradually getting harder to find (Bloom et al., 2020), which says that no matter how much resources (including scientists, R&D capital, and AI) we will throw at R&D in general, and developing AGI in particular, we just won't ever get to AGI. I don't believe in this scenario primarily because, looking at the millennia of past technological development, I observe that the rates of technological progress have been systematically *increasing*. With each new era progress accelerated by at least an order of magnitude. And although there have been temporary setbacks, and there may well be more such setbacks in the future (for example if we realize that further progress in AI software requires quantum computing and we'd have to accumulate the stock of quantum computing hardware from scratch to do that), it just doesn't seem plausible to imagine that technological development—which has never been so fast as in the digital era—would just *stop forever* without any external reason.

Second, we may encounter such an external reason—a major disaster that would permanently curtail the development potential of the human civilization before the arrival of AGI. This is a doomsday scenario in itself. Think, for example, of a global nuclear war followed by a nuclear winter, a major asteroid impact, depletion of an *essential*, non-replaceable natural resource (if there is such thing at all), a global pandemic (natural or man-made), or the like. In this regard I would say: yes, there are other existential threats to the humankind other than AGI and it's important to keep them in check to the greatest possible extent. However their combined probability is—as far as Nick Bostrom and his colleagues can see (Bostrom, 2002)—very low, especially at the timescale of just a few decades that I'm considering here in the context of arrival of AGI.

Finally, the humankind may be successfully prevented or otherwise lose interest in developing AGI. Perhaps there would be a globally coordinated ban on AI research and its funding (a permanent "AI winter"), a successful global anti-AGI campaign that would convince people that developing AI with broad applications is evil, or a global dictatorship which would outlaw AI. Again, I don't believe in any of that in the considered timeframe. Crucially, as the case of CO_2 emissions and global warming has already shown, universal coordination is notoriously hard to achieve when there are strong individual incentives to deviate. And gains from developing advanced AI are broadly acknowledged. If anything, the dynamic we're seeing now is not of coordinated action against AGI but an *arms race toward AGI*. Moreover, it's not at all clear whether slowing down AI progress is good policy, given the evidence we have that well-designed AI algorithms short of AGI can actually do a lot of good for the human society—think for example about the massive reduction in road fatalities which may be achieved thanks to autonomous vehicles, or the massive improvement in our interpersonal communication abilities possible to obtain thanks to automated real-time language translation.

Failure in designing friendly AGI. This is probably the most important condition because, unlike the former two, here everything is really in our hands. Nevertheless, there are a few reasons to be anxious about the final outcome.

First, literally *millions* of different things can go wrong with AGI. "Anna Karenina principle", coined by Lev Tolstoy to describe unhappy families ("All happy families are alike; each unhappy family is unhappy in its own way") and used by Jared Diamond to describe why so many animals are non-domesticable (Diamond, 1997), applies just as well to the hypothetical failed AGI attempts. There is just one way to set its objectives right and millions of ways to do it wrong. Even the most well-intentioned attempts will likely fail to secure that the AGI will be friendly because our understanding of the world is limited by our cognitive powers and we can't comprehend what new dimensions the AGI may come up with to achieve its goal. "Make everyone happy"—and we'd be permanently drugged with heroin. "Make everyone happy without using chemical substances"—and we'd have electrodes installed in our brains which would make us feel as if we were drugged with heroin. "Make the world such that no human being will ever suffer"—and we're all dead: lack of people is the safest precondition for the lack of human suffering. And so on. For any objective function of the AGI conceived so far, there is always something horrific that follows. Eliezer Yudkowsky from the Machine Intelligence Research Institute (MIRI) bets that for any closed-form objective function that you may come up with, he'll tell you why it certainly won't benefit the humankind. The system of human values is complex and we're far from understanding it ourselves.

Second, we don't even have a clear compass telling us where we should be heading. There is no universal set of human values that could be instilled in the AGI. The values people claim to adhere to vary wildly across cultures and time. Moreover, any single person can fall victim to some form of personal enlightenment, revelation or brainwashing, and change her view on these things in a dramatic manner. And most importantly, the humankind as a collective gradually refines its morals, its worldviews, and its values in reaction to the progress in science, technology, and economic growth (Harris, 2010). If medieval scholasticists somehow created AGI and instilled their well-intentioned but fundamentally medieval values in it, modern people wouldn't be too happy with the outcome, would we? (Bostrom and Yudkowsky, 2011).

The bottom line is that we're getting closer and closer to AGI but we still don't know what we should wish for it to do. We know that there is no simple solution to the conundrum and that our value system is probably complex. We also know that restrictions, boundary conditions and behavioral laws won't work because a sufficiently intelligent entity would always find a way around them. We are aware of the mechanism of instrumental convergence so that no matter what AGI goal structure we come up with, its pursuit will likely result in self-protection, replication, recursive self-improvement, and capture of useful resources. Therefore the only way toward friendly AGI is probably through gradual learning. Instead of coding our values in the AGI in a verbatim way, researchers would have to teach them in some form of a supervised learning scheme with correction mechanisms to intervene when things begin to go wrong. However, perfect alignment of AGI goals with human values can only be achieved—if at all—when the AGI learns our *true* values from our actual actions and decisions rather than declarations and simplified theories. Given that our behaviors and moral codes evolved greatly over the last centuries, the AGI should also learn to extrapolate their evolution beyond the observable horizon, taking advantage of its superior intelligence. And the structure of learned goals must be stable with respect to the discovery of new dimensions of action and under recursive re-writes of the algorithm in the quest for self perfection. And it should also prohibit sub-goals from replacing the final goals. That's awful lot to learn.

The biggest worry is, however, that we may have just one shot in creating friendly AGI (Hanson and Yudkowsky, 2013). This is bad news because historically our progress always relied on long sequences of trial and error, controlled systematic improvements of the initial flawed but promising idea. We have tried and failed repeatedly, but finally we managed to nail it and succeed. This happened both with new technologies (think of zeppelins as means of aerial travel, later replaced by airplanes), new scientific theories (think of phlogiston physics, the geocentric model or the miasma theory of disease, replaced respectively by the oxygen theory of combustion, the heliocentric model and the germ theory), and also with handling major catastrophic risks. The humankind had to go through Hiroshima and Nagasaki to learn that we'd better not use nuclear weapons against each other anymore, and through large-scale lead poisoning to understand that our cars should rather run on unleaded fuel. In the case of AGI, however, from the moment it undergoes intelligence explosion there will be no turning back. We'd better get it right the first time.

Unfortunately, premature launch of unsafe AGI is quite likely. The reason is that while each human individual tends to be risk-averse and avoid lotteries with potentially big losses, this regularity doesn't apply to our species as a whole. The collective human brain is apparently a risk lover: initiatives that bring potentially both large gains and losses will rather be undertaken than not. This is because with increasing returns to scale, it's not the actions of the average human that matter the most for aggregate development but the actions of the select few who are either most inclined to take risks or most biased in their positive assessment of the outcome. And there are also additional amplifying mechanisms. For example, in many cases the gains can be appropriated while the losses are diffuse—think of oil proceeds and greenhouse gas emissions, or of the huge profits of few successful startups and relatively harmless losses of the countless failed ones. Furthermore, there is also the *unileralist's curse*. Suppose that "each of a number of agents (...) decides whether or not to undertake [an initiative] X on the basis of her own independent judgment of the value of X (...) Each agent's judgment is subject to error—some agents might overestimate the value of X, others might underestimate it. If the true value of X is negative, then the larger the number of agents, the greater the chances that at least one agent will overestimate X sufficiently to make the value of X seem positive." Bostrom et al. (2016, p. 351) The unilateralist's curse implies a high probability that when there will be, say, 20 AI researchers contemplating whether a given AGI design is safe, with nineteen of them saying it's not, the remaining one may have a different opinion and unilaterally start the algorithm. To mitigate this, (Bostrom et al., 2016) argue for a *principle of conformity* which could reinstate risk aversion in the collective human brain and reduce the risk of harmful unilateral action.

11.5 Human Civilization Without Humans?

To understand how the world may look like after AGI undergoes intelligence explosion, it's useful to make the distinction between the survival of humans as a biological species and survival of the human civilization. After all, if the existential threat from AGI materializes these two may no longer coincide.

So what would happen if the doomsday scenario materializes and AGI makes humans extinct? Certainly the world won't become an empty desert, but rather *a rapidly developing, human-less but human-enabled civilization* that is certainly *after something*. The question is then: what would that *something* be, and could today's humans be potentially proud of it? Would it be a "worthy" continuation of the human civilization, say, one that discovers ever newer laws of nature and puts them into productive use, creates an increasing variety of increasingly complex objects, transcends its limits and colonizes the surrounding universe? Or would it be something deeply disappointing, like an AGI blindly turning matter and energy into an endless deluge of paperclips? (Bostrom and Yudkowsky, 2011; Bostrom, 2014) Can we make sure that the extent of control of Earth-based intelligence will continue to grow and generate a broad array of achievements?

The key problem with making such evaluations is that we haven't figured out the purpose and value of human life yet, both of each individual oneself and of the whole society combined. In this book I argue a lot for local control maximization as the proximate purpose of our lives, but—to reiterate—it's just a collection of instrumental goals that are consistent with a wide range of final goals and not the final goal itself. A term like "a worthy continuation of the human civilization" is then necessarily vague. (In fact, we can't even say with certainty that *our own* civilization is a worthy continuation of itself. For example, is the industrial era a worthy continuation of the agricultural era? Are the Middle Ages a worthy continuation of Antiquity?) What we could do so far was to trace the overall extent of our control across centuries, using proxy measures such as total population, world GDP, energy consumption, the volume of communicated data, or the average level of self-reported happiness among the *homo sapiens*. In this narrow sense we can say *ex post* that we're indeed on the right track and our civilization is developing fast. But it's also rather clear that these proxy indicators are not goals in themselves and treating them as final goals for the future would quickly lead us astray. Especially in the face of AGI.

Some major questions remain in place even if the AGI were able to learn the true human values perfectly, though. Think in particular about the fact that our behaviors and moral codes evolved greatly over the last centuries. For example, slavery has been abolished only very recently (e.g. in the nineteenth century in the USA), and the struggle against racial, ethnic, gender and religious discrimination is still continuing well in to the twenty-first century (in some parts of the world, it has hardly started). Would the AGI categorize the observed progress in rights equalization and social inclusion as part of a trend toward ever deeper wisdom, changes that should have happened much earlier and in any circumstances, or an economically motivated phenomenon, marking the increasing demand for human cognitive work in the industrial and early digital era and the concurrent rise in global economic and social interconnectedness? Would it view this change as permanent or reversible? (And what would *we* think about this dilemma ourselves when most people would have become economically useless amidst advancing automation? Would then the "still useful human elite" be able to resist the temptation to discriminate? And would the *homo sapiens* trust the AGI in its assessment of such fundamental issues?)

Furthermore, it remains a mystery how the humans would behave in a world where our needs can be easily satisfied, consumption goods and leisure are abundant, but the environment becomes fundamentally incomprehensible and all important decisions are made on our behalf by the AGI? Would *observing* progress be enough or would we need to feel that we're *causing* it? And would there be education as we know it once human skills become obsolete for the labor market? Crucially, can a world where humans and AGI coexist be stable, or is it doomed to collapse so that the superior AGI would have to eliminate us anyway, maybe even in self-defence?

This section has so many question marks because complex phenomena such as advanced AI are harder to understand than to set in motion. In retrospect we know that the development of the human civilization has brought all these mounting side effects but we couldn't predict them in advance. This is also why we're so lost when trying to set our expectations regarding the coming superhuman AGI: its properties won't be reducible to anything we could work out and fully understand in advance—or otherwise it wouldn't be superhuman!—and we may not have a second try to fix the undesired outcomes. Our *ex ante* understanding of the AGI and the civilization it will build is going to be, at best, the same as our understanding of the human brain or the financial market today: we understand the principal building blocks but not the emergent complexity. And after the AGI undergoes spontaneous intelligence explosion, even the building blocks may become hard to understand.

11.6 Transhumanism Is Modesty

So far I've maintained a sharp distinction between human and artificial intelligence. I considered the consequences of AI capabilities taking off relative to the benchmark of human intelligence which I depicted as more or less fixed. But what if this dichotomy is false, and in the future humans and AI will *blend*, giving rise to hybrids of human minds and networked digital computers, with far superior cognitive skills but unchanged human goals? What would you feel about such a scenario? Are you a *transhumanist*?

To wrap one's mind around the concept of human brain augmentation, it's instructive to see it already happening. Digital devices help our thought processes on a daily basis: Internet with rich data sources and efficient web search engines help us collect, recall and organize information, computers with computational software help us do the math and create predictions, smartphones with GPS tracking help us navigate the world. Not to mention good old inventions like radio and telephone which help us communicate information across long distances. All these technologies facilitate our cognitive work and make it more efficient.

The problem is, though, that such devices can help us only to the point we're able to absorb and process their help. One has to have the prerequisite skills to use the available technologies appropriately. "In a world where bits travel at thousands of miles per second, the last 20 inches of the journey—those separating the screen from the user—are where communication stops," says Cesar Hidalgo. Our limited cognitive bandwidth has by now become a key obstacle in the development of broadly understood software and the crucial bottleneck of economic growth. It's also the key reason why greatest economic returns from automation are observed only when processes are fully automated.

The alternative to replacing humans with AI algorithms with superior data processing power and communication bandwidth is to increase the human–computer communication bandwidth by integrating digital hardware directly into the human brain, thus bypassing eyes and ears as our data input, and speech and writing as our data output. This could potentially be done, for example, by creating an artificial "outer layer" for our biological brain, similarly to the way our cortex is an outer layer for the evolutionarily older reptilian and paleomammalian brain. This kind of technology has been envisaged, for example, by the restless visionary Elon Musk and other co-founders of Neuralink (though Elon Musk only does this in his spare time when he's not currently thinking about autonomous vehicles by Tesla or flying to Mars with SpaceX). The potential of such technology is yet unknown; nevertheless—if feasible—it clearly gives hope for increasing human intelligence (or more broadly, cognitive skills) in the future. And that would be strongly needed to make humans a more worthy competitor for rapidly advancing AI algorithms.

Even more curious is the idea of *mind uploads*, whole brain emulations. This hypothetical technology would allow us to scan our entire brains to the level of each dendrite, represent them in a digital form, and run on a computer with far superior processing speed and communication bandwidth (Hanson, 2016b). In this way, the pace of growth in human cognitive work at the global scale could be vastly accelerated and eventually tied to growth in programmable hardware rather than the human population. Mind uploading seems to be also a potential shelter against a rising unfriendly AGI: it may potentially allow us to at least save our software—our minds—from being wiped out together with our bodies.

A big philosophical question coming with the idea of mind uploads is, however, would the uploads still be *us*? Would they carry our consciousness and sense of continuity? And what about the technical possibility to run the brain process simultaneously on many devices and terminate at will? Is stopping a brain emulation process the same as killing a human mind? Is slowing it down an act of discrimination? (Hanson, 2016b)

The bottom line is that the digital era sets up a major challenge for our human cognitive skills. If our brains remain internally unaugmented, not only may some of us eventually become economically useless, but even the entire collective human brain may lose its primacy as the most intelligent entity on Earth and the most potent local control maximizer. Transhumanism recognizes this challenge and proposes to move ahead by joining forces with the rapidly growing computer hardware and AI algorithms.

I perceive transhumanism as a fundamentally modest approach. It rejects the fantasy of the human mind being somehow magically unique and automatically superior to all other conceivable minds. It aligns with the scientific evidence suggesting that our brains are not qualitatively but only quantitatively different from certain animal brains, particularly of other primates. It also accepts that there is no physical constraint precluding artificial brains from being more efficient than the human brain, and in fact there is evidence that our cognitive skills may be limited by certain hardware requirements. The brain must fit inside the cranium, and cranium size is limited by the anatomy of the pelvis of a woman giving birth. The brain is also a heavy burden to our energy budget: an adult human expenses about 20% of daily energy on the functioning of her brain, even though it constitutes on average only 2% of her mass. Unconstrained by such factors, human brains could likely benefit from number of improvements, from an enlarged frontal cortex to a wi-fi connection to the Internet. Transhumanism also rejects *carbon chauvinism*, accepting that both *sentience* and *sapience* (both consciousness and high intelligence) can be reached also by minds made of a different substrate than proteins (Bostrom and Yudkowsky, 2011). It doesn't discriminate between information stored, processed and communicated within hardware made of various substrates, be it carbon or silicon, neurotransmitters or semi-conductors.

Let me also make two final remarks here. First, there is much more to transhumanism than brain–machine interfaces and mind uploads. Humans can be augmented through numerous other means, e.g. by selective alterations of the human genome. Human DNA edits are apparently already coming: the Chinese biophysicist He Jiankui has reportedly successfully edited the genome of a single-cell embryo with CRISPR/Cas9 technology leading to the birth of twin girls with genetic HIV immunity (for which, though, he was fired, fined, and sentenced to jail). One of the Holy Grails of transhumanism is to stop the natural aging process, extending healthy human life into an arbitrary long time (see Nick Bostrom's fable of the dragon-tyrant, Bostrom (2005)).

Second, even bold transhumanist solutions may not be enough to win the race against AGI. In particular, we can't exclude the possibility that certain fundamental properties of the human brain are an unsurmountable liability and intelligence explosion is much easier to obtain with a fully artificial AI structure. For example, even if brain–machine interfaces equate the bandwidth of human–human and human–computer interaction to the bandwidth of our internal thinking, it may still turn out way slower than computer–computer broadband data transmission. Transhumanist solutions promise a major boost to the human brain in our "race against the machine", but whether this boost would be enough to win the race—we really don't know. (By the way, see that I'm being a Militant Agnostic here: "I don't know but you don't either".)

11.7 Selected Alternative Scenarios

It seems that we're heading full speed toward AGI which is going to be an ultimate knife edge for human control and the flourishing of the human civilization. The AGI holds both enormous promise and existential threat, and begs us to discuss and redefine such foundational ideas like the "human mind" and "human values". What used to be philosophical problems with no immediate practical implications, are now becoming urgent engineering problems. As Nick Bostrom once remarked, we're now in pressing need of "philosophy with a deadline".

There are too many unknowns, however, to say that this kind of future is our certain destiny. Therefore in this section I outline three alternative scenarios. I am discussing these scenarios only briefly, though, because they seem to be somewhat less likely than the AGI intelligence explosion ("hard AI takeoff") scenario that I elaborated at greater length. My expectations in this regard are based on extrapolating past exponential trends in scientific achievement, output complexity, computer hardware and software accumulation (Kurzweil, 2005; Hilbert and López, 2011; Grace, 2013; Hidalgo, 2015; Hernandez and Brown, 2020; Davidson, 2021), as well as qualitative economic trends related to the progress in automation, business scalability and winner-takes-all market dynamics in the digital sector (Andrews et al., 2016; Autor et al., 2017; Brynjolfsson et al., 2019; WID world, 2018). I ground my reasoning in the hardware–software model and presume that humans will continue to pursue local control in the same way they did in the past. Nevertheless, like any extrapolation, my predictions may go terribly wrong. So it's good to know the alternatives.

Figure 11.1 locates the hard AI takeoff scenario as well as three alternative ones along a single vertical axis that represents the returns to AI cognitive reinvestment (Yudkowsky, 2013) or equivalently, the maximum potentially achievable level of artificial intelligence. These scenarios are: (i) secular stagnation or non-accelerating R&D-based growth, (ii) race against the machine, and (iii) human-level AI (possibly through whole brain emulation).

Secular stagnation or R&D-based growth. The most conservative scenario assumes that we already have or will soon reach the limits of AI development and that AI will not play an important role in further economic growth. In this vein, based on *recent* economic trends (using, e.g., postwar data for the developed economies) in GDP per worker and residually computed total factor productivity (TFP) economic scholars like Fernald, 2015, Gordon, 2016, Bloom et. al., 2020 find that in the twenty-first century GDP and TFP growth is slowing down, and hypothesize that it will slow down even further in the future as real convergence of catching-up countries advances and global population growth diminishes.

Fig. 11.1 Maximum AI intelligence versus human intelligence

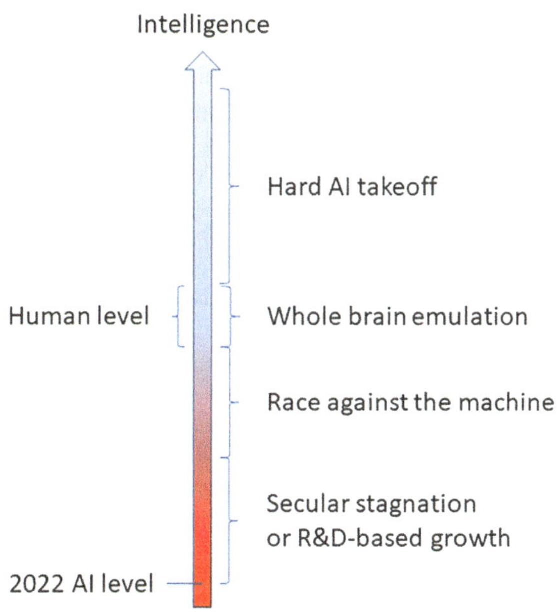

This line of economic thought assumes that human cognitive work will always be an essential input to production and R&D. It reduces the digital revolution to its hardware component, and is keen on drawing analogies between the digital and the industrial revolution. Specifically, recent technological developments are sometimes characterized there as the "fourth industrial revolution" (Schwab, 2016)—where the steam engine is the first industrial revolution, electricity is the second, computers are the third, and Internet (plus a broad array of cyber-physical systems) is the fourth. Unfortunately, this line of thought fails to acknowledge the important software component of the digital revolution. It doesn't recognize that the key achievements of the "third and fourth industrial revolution" were in fact not in the realm of physical action, but information processing. Consequently, when viewed through the lens of the hardware–software model, the secular stagnation scenario requires that limits to AI development must be prohibitively low and the scope of further automation must be firmly limited. Only then will human cognitive work forever remain the key bottleneck of economic growth and determine its long-run pace.

Would it be good for us if this scenario materializes? Yes and no. The downside is that the tremendous growth in data processing and communication that we're observing these days will never translate into similarly fast growth in GDP, human flourishing and human control. The upside is, in turn, that it eliminates AGI from the list of our existential threats. A popular view among economists, politicians and public opinion alike is that secular stagnation is a *pessimistic* scenario that warrants steps to be taken to avoid it. Apparently, twenty-first century levels of human prosperity proceeding into indefinite future are *not enough* for us, and the existential threat from AGI remains under the radar.

This scenario comes in two flavors, depending on the extent of scale effects in R&D (Jones, 1999), that is, the extent to which consecutive ideas are getting easier to find, and the role of R&D capital (Growiec, 2022b). In the case of weak (less than proportional) or even negative scale effects and a bounded stock of R&D capital, the prediction is of a *secular stagnation*: in the very long run, in the absence of population growth, there is gradual convergence to a steady state. Economic growth and technological progress slow down until they completely stop. By contrast, in the case of strong (proportional) scale effects, or weak scale effects combined with unbounded accumulation of R&D capital, we would experience perpetual exponential *R&D-based growth*: GDP grows because of the systematic development of technologies augmenting human cognitive work, and R&D keeps its pace in developing these technologies because of the proportional feedback from old to new ideas, either directly or via the accumulation of R&D capital. But there is no growth acceleration.

Race against the machine. The second alternative scenario assumes that on the one hand there is ample potential for additional automation and AI development in the coming decades, but on the other hand the humankind will never figure out how to create software that could approach human-level intelligence. This scenario expects a lot of disruption in the economy due to automation, but also predicts that humans will always remain at least one step ahead of AI algorithms in our "race against the machine". (Perhaps the proponents of this scenario believe that superhuman intelligence just cannot follow from human design? Indeed, neuroscientists have long been saying that our brain is a catch-22: "If the human brain were so simple that we could understand it, we would be so simple that we couldn't". However, this seeming paradox excludes the possibility of our cognition being *augmented*, which is already happening.)

Excellent analysis of this scenario has been carried out by Acemoglu and Restrepo (2018). In their model, existing tasks are gradually automated—and there is no task that would be non-automatable on principle—but humans stay ahead of AI and make themselves useful by gradually developing new tasks that are at first not automated. AI, in contrast, doesn't invent any new tasks. There are three distinct ways in which such an economy may unfold over the long run: the share of automated tasks may either grow toward 100%, stabilize below 100%, or gradually fall to zero. In the last case, automation eventually becomes irrelevant, which brings us to the secular stagnation / R&D-based growth scenario that I have already covered. Otherwise, the human future is generally bright: there is perpetual economic growth, driven by R&D and strongly accelerated by automation, and there is no existential risk from AGI because AGI is infeasible. There are, nevertheless, two worrying trends to observe in this scenario (even though they are not explicitly included in the Acemoglu–Restrepo model in which all people are alike). First, as each and every task will eventually be automated, people will be constantly pushed to learn new skills and move to new not-yet-automated occupations. Second, as the complexity and skill requirements for the newly invented, not-yet-automated tasks will be constantly growing, there will be also an increasing group of people who wouldn't catch up, while the remaining—increasingly valuable—cognitive work will be provided by the ever fewer others. Therefore ongoing automation will increase demand for scarce specialized skills, which would naturally elevate income inequality.

Moreover, as this scenario expects that life-threatening AGI won't be created, but our economy will nevertheless increasingly rely on software services of sophisticated AI algorithms, it will be the perfect environment for developing ever more proficient algorithms that would *hack human minds* (Harari, 2017), i.e., perform subtle manipulations of thought processes in our brains to coerce us into doing things we wouldn't otherwise like to do. (In the presence of an AGI, our actions would be of no major importance anyway, so there would be no point in manipulating them. Therefore this point is important only in the absence, or prior to full-blown AGI.) In consequence, this scenario carries important political connotations: it may potentially undermine democracy, leading to a persistent digital dictatorship.

Finally, both secular stagnation / R&D-based growth and the "race against the machine" scenario downplay the role of any future technological revolutions in hardware, such as nanotechnology, quantum computing, fusion power or advanced bioengineering. Both of them say that no matter how advanced our hardware would be, human cognitive work will always remain the limiting factor of further development. And no matter how great our AI-augmented cognition would be, we will never sufficiently understand the emergent properties of the human brain to be able to simulate it on a computer, creating human-level AI.

Human-level AI. In the third alternative scenario the maximum possible intelligence of AI algorithms is at par with the human. This scenario doesn't place limits on human ingenuity, but assumes instead that the human brain is somehow an *optimal structure* that cannot be improved upon or scaled up to gain qualitative cognitive improvements—it can only be potentially sped up. There are intuitions to the contrary—e.g. cortex size has been shown to correlate with cognitive abilities in mammals, and our cortex size is limited by the size of our skulls, which is in turn capped by the size of pelvis of women giving birth—but let us proceed with this assumption nevertheless. (Also, this means that not only the *level* of intelligence but also the *properties* of the AI will be analogous to those of the human brain. The case where the AI would be completely alien but somehow similar to us in the level of intelligence seems insanely unlikely and can be effectively lumped either with the "race against the machine" or the hard AI take-off scenario. Returns to intelligence appear to be so big that only the single most intelligent entity in the environment can effectively exert local control (Good, 1965).)

This is a transhumanist scenario where biological and emulated human brains coexist and potentially cooperate in performing cognitive actions (Hanson, 2016b). From the economic perspective, an important implication is that human cognitive work would be then decoupled from the physical human population, and instead become proportional to programmable hardware. If computing power, data storage and communication would continue to grow at Moore's Law pace that we've witnessed in the recent decades, that would imply a massive acceleration in economic growth. Freed from the limiting factor of slowly growing software capacity, global GDP could double perhaps every three years or so, until Moore's Law dies down or a new hardware revolution comes along, increasing the pace of hardware accumulation by an additional order of magnitude. In the (rather quickly approached) limit, all tasks will be either conventionally automated or outsourced to the emulated human brains; the embarrassingly slow cognitive work of biological humans will become redundant.

When discussing such a scenario, it must be mentioned that the more advanced the hypothetical AI is, the more it is loaded with moral dilemmas. Human-level AI, in particular, seems to deserve human moral status (Bostrom and Yudkowsky, 2011). But should we really treat the needs of emulated brain processes equally with needs of physical humans? That's a tough question because the "demographic explosion" of emulated brains which scale proportionally to programmable hardware may likely place biological humans under existential threat in a Malthusian race for resources (Korinek and Stiglitz, 2019). Furthermore, even if complexity and depth of thinking in biological and emulated brains will be equal, the pace of data communication and processing in the digital emulations may be *overwhelmingly* faster (and accordingly greater would be the energy consumption). Given a tremendous difference in the perceived rate of time, wouldn't this make the needs of emulated brains a higher priority? If so, then this scenario becomes very analogous to the hard AI take-off scenario. And the bad news is that the goal of the emulated brains will be the same as our human goals. They will maximize local control just like we do, and that *certainly* places them in the "unfriendly AI" territory.

11.8 How Long Is the Long Run?

Planning for the future, it's important to set priorities. Reviewing the challenges for the digital era, I mentioned the ecological consequences of our increasing local control (particularly climate change), psychological consequences of the data explosion, increasing inequality in the digital-era economy, growing scale mismatch between the economy and policy, job displacement due to automation, increasing strain on representative democracy, distortion of our choices by AI algorithms hacking our minds, and the potential rise of AGI. Which of those challenges are most important? Which are most urgent, and which can be dealt with later? And what's the time frame we're talking about—five years, twenty, fifty, hundred, or more?

Well, I can't tell you exactly. Setting priorities over such profound issues requires deep thought and broad debate. Odds are, however, that these problems are *too important* and *too urgent* to be dealt with competently. None of them can be solved single-handedly (even if Elon Musk tries hard) or by the means of local or national political agendas. They are global problems which require global solutions. If that's the case, though, unfortunately the most human thing to do is to wave our hands, say "be it as it may", and return to our daily routines. That's why—I predict—our policy reaction to the unforeseen consequences of digital technologies will be similar to the current policy reaction to climate change: too weak and too late. For the same reason I feel confident when predicting that the secular trends seen in historical data will continue into the future, no matter how troubling they are.

Which doesn't mean we *should* wave our hands and move on. To the contrary. In such an intense time, packed with so many technological opportunities and risks, rational prioritizing is key. I have four suggestions.

First and foremost, never ever should we underestimate existential risks. Existential risks are, like, *infinitely* bad—so bad, that they should be avoided at *all* costs. Just look at all the fascinating human history and imagine it suddenly ends. Forever. Figure 11.2, which I borrowed from Nick Bostrom, is one of those figures that are literally worth a thousand words. If you study it, you'll see why I'm discussing the AGI issue—the only true existential threat among the lot—at such great length in comparison to other long-run issues of the digital era, such as growing income inequality or the risk of AIs hacking human minds.

Second, the time frame for these "long-run issues" is actually worryingly short. Numerous effects of the digital era—from massively positive, like the digital augmentation of our cognition, democratization of information creation and access, or substantial cost cuts thanks to establishing global value chains, to frustratingly negative, like concentration of benefits from economic growth among the top 1%—are affecting us already and will intensify in the future. And then, according to many experts (from AI researchers to physicists—but, as far as I know, generally excluding economists) there may be a technological singularity. Experts predict that AGI may emerge some time around 2050 (with a wide range of uncertainty, though). Chances are that it's going to happen within your lifetime, and certainly in the lifetime of your children. In the strict economic sense (Comin and Gertler, 2006) it's not

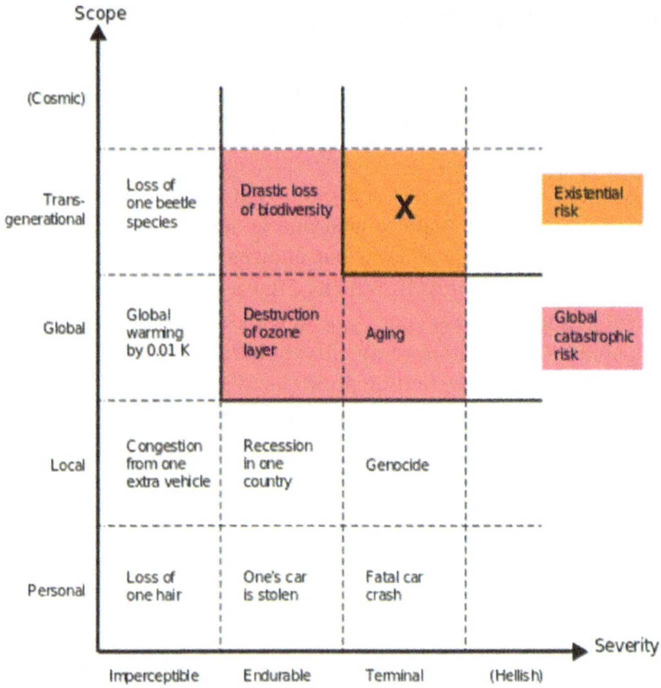

Fig. 11.2 Existential risks viewed against other kinds or risks. *Source* Bostrom (2002)

even long run, it's *medium run* (8 – 50 years)! From the technological singularity onwards, it's going to be either a bonanza of human control—we may, for example, stop the aging process, develop new fantastic ways of obtaining energy, and colonize space—or sudden death of the humankind. The ultimate knife edge, the next *filter*, is approaching fast. Let's not be surprised.

Third, we shouldn't underestimate the run-up to the singularity. The path we'll take toward AGI may determine not only the timing of its arrival, but crucially also the likelihood of a favorable outcome. Cooperative action in the area of AI research—think of open access to research papers, data, and free open source software—appears to be a more robust framework than secretive R&D efforts amidst an AI arms race of software firms or even countries (e.g. USA versus China). Adoption of the open access framework will also allow new narrow AI algorithms (short of AGI) to diffuse more broadly and benefit more people. It is also crucial that AI safety research is treated as priority and common safety standards universally implemented.

Fourth, both existing problems like global warming and tax evasion by global enterprises of the rapidly expanding digital-era economy, and future challenges like ensuring AGI safety, require globally coordinated policy. Even best-intentioned actions, like our climate agreements from Kyoto and Paris, will get nowhere without universal compliance and support. When it comes to solving global problems and avoiding existential risks, global coordination is key. Naturally, there is no guarantee of success because someone may always unilaterally decide to perform globally harmful action; for example launch a nuclear missile or an unfriendly AGI, but we also know that pursuing narrow national interests in the digital era, with its highly scalable technologies and massive data flows that don't respect national boundaries, is a recipe for failure.

* * *

Now let's take one more step back and frame the human history within the Big History of the whole physical universe (Carroll, 2016). At this grand scale, the distinctive feature of life on Earth, and the humankind in particular, is that it excels in building complexity. Our networked brains and their products—the global economy, science and the Internet—are the most complex structures that have ever existed in this corner of the universe. And as far as we can tell, this complexity is growing explosively, at ever increasing exponential rates.

However, the human civilization, like all matter and energy in the universe, is constrained by the second law of thermodynamics which states that the overall entropy (disorder) of the universe is systematically going up. How does this square with the fact that growing complexity requires increasing order?

The key point is that entropy and complexity aren't really opposites. As Carroll (2016) explains, over the course of approximately 13.8 billion years since the Big Bang the universe is gradually evolving from a low-entropy, low-complexity primordial soup to a high-entropy, low-complexity empty space end state. Only in the meantime and in local, non-isolated systems, like the Earth awash with photons of low-entropy energy from the Sun, can complexity increase. Complexity is built, though, and the necessary information is stored, communicated and processed, only at the cost of expending energy, that is—as physicists would say—transforming useful low-entropy energy into less useful high-entropy energy. This complexity may come in the form of biological cells, multi-cellular organisms, but also humans and their sophisticated products, or—perhaps soon—superintelligent AI.

Eventually, complexity must fall apart and give way to disorder. Even if in our own eyes the human civilization is a big boom, in the grand scheme of things it is just a temporary and local uptick in complexity. For what we know, however, it's entirely on us—the collective human brain—how high this uptick (or boom) will be, how long it will last, and to which extent it will reflect the ideas and actions of our species. There is no clear upper bound for complexity, but to keep increasing it our civilization will have to pass a few more filters, the first one looming some time around the 2050s. With development and scale, stakes are going up. The contemporary digital era may be either a new beginning or beginning of an end.

References

Acemoglu, D., & Restrepo, P. (2018). The race between man and machine: Implications of technology for growth, factor shares and employment. *American Economic Review,108*, 1488–1542.

Andrews, D., Criscuolo, C., & Gal, P. N. (2016). *The global productivity slowdown, technology divergence and public policy: A firm level perspective* (Working Party No. 1 on Macroeconomic and Structural Policy Analysis). OECD.

Autor, D., Dorn, D., Katz, L. F., Patterson, C., Van Reenen, J. (2017). *The fall of the labor share and the rise of superstar firms* (Working Paper No. 23396). NBER.

Bloom, N., Jones, C. I., Van Reenen, J., & Webb, M. (2020). Are ideas getting harder to find? *American Economic Review,110*(4), 1104–1144.

Bostrom, N. (2002). Existential risks: Analyzing human extinction scenarios and related hazards. *Journal of Evolution and Technology, 9.*

Bostrom, N. (2005). The fable of the dragon-tyrant. *Journal of Medical Ethics,31*, 273–277.

Bostrom, N. (2012). The superintelligent will: Motivation and instrumental rationality in advanced artificial agents. *Minds and Machines,22*, 71–85.

Bostrom, N. (2014). *Superintelligence: Paths, dangers, strategies*. Oxford University Press.

Bostrom, N., & Yudkowsky, E. (2011). The ethics of artificial intelligence. In W. Ramsey & K. Frankish (Eds.), *Cambridge handbook of artificial intelligence*. Cambridge University Press.

Bostrom, N., Douglas, T., & Sandberg, A. (2016). The unilateralist's curse and the case for a principle of conformity. *Social Epistemology,30*, 350–371.

Brynjolfsson, E., Rock, D., & Syverson, C. (2019). Artificial intelligence and the modern productivity paradox: A clash of expectations and statistics. In A. Agrawal, J. S. Gans & A. Goldfarb (Eds.), *The economics of artificial intelligence: An agenda*. University of Chicago Press.

Carroll, S. M. (2016). *The big picture: On the origins of life, meaning, and the universe itself.* Dutton.

Chalmers, D. (1995). Facing up to the problem of consciousness. *Journal of Consciousness Studies,2*, 200–219.

Comin, D., & Gertler, M. (2006). Medium-term business cycles. *American Economic Review,96*(3), 523–551.

Davidson, T. (2021). *Could advanced AI drive explosive economic growth? Report* Open Philanthropy.

Dennett, D. (2017). *From bacteria to Bach and back: The evolution of minds*. Norton & Company.

Diamond, J. (1997). *Guns, germs, and steel: The fates of human societies*. Norton & Company.

Etzioni, O. (2016). No, the experts don't think superintelligent AI is a threat to humanity. *MIT Technology Review*.

Fernald, J. G. (2015). Productivity and potential output before, during, and after the Great Recession. *NBER Macroeconomics Annual,29*, 1–51.

Good, I. J. (1965). Speculations concerning the first ultraintelligent machine. *Advances in Computers,6*, 31–88.

Gordon, R. J. (2016). *The rise and fall of American growth: The U.S. standard of living since the civil war*. Princeton University Press.

Grace, K. (2013). *Algorithmic progress in six domains* (Technical report 2013–3). Machine Intelligence Research Institute.

Grace, K., Salvatier, J., Dafoe, A., Zhang, B., & Evans, O. (2017). *When will AI exceed human performance? evidence from AI experts* (Working paper). Future of Humanity Institute, Oxford University.

Growiec, J. (2022b). What will drive global economic growth in the digital age? *Studies in Nonlinear Dynamics and Econometrics*, forthcoming.

Hanson, R. (2016b). *The age of Em: Work, love and life when robots rule the earth*. Oxford University Press.

Hanson, R., & Yudkowsky, E. (2003b). Ethical issues in advanced artificial intelligence. In G. E. Lasker, W. Wallach & I. Smit (Eds.), *Cognitive, emotive and ethical aspects of decision making in humans and in artificial intelligence* (Vol. 2, pp. 12–17). International Institute of Advanced Studies in Systems Research and Cybernetics.

Hanson, R., & Yudkowsky, E. (2013). *The Hanson-Yudkowsky AI-Foom Debate*. Machine Intelligence Research Institute.

Harari, Y. N. (2017). *Homo deus: A brief history of tomorrow*. Vintage.

Harris, S. (2010). *The moral landscape: How science can determine human values*. Free Press.

Hernandez, D., & Brown, T. B. (2020). *Measuring the algorithmic efficiency of neural networks*. arxiv:2005.04305, arXiv.

Hidalgo, C. (2015). *Why information grows: The evolution of order from atoms to economies*. Basic Books.

Hilbert, M., & López, P. (2011). The world's technological capacity to store, communicate, and compute information. *Science,332,* 60–65.

Jones, C. I. (1999). Growth: With or without scale effects? *American Economic Review,89*(2), 139–144.

Korinek, A., & Stiglitz, J. (2019). Artificial intelligence and its implications for income distribution and unemployment. In A. Agrawal, J. S. Gans & A. Goldfarb (Eds.), *The economics of artificial intelligence: An agenda*. University of Chicago Press.

Kurzweil, R. (2005). *The singularity is near*. Penguin.

Muehlhauser, L., & Salamon, A. (2012). Intelligence explosion: Evidence and import. In A. Eden, J. Søraker, J. H. Moor, & E. Steinhart (Eds.), *Singularity hypotheses: A scientific and philosophical assessment*. Springer.

Omohundro, S. M. (2008). The basic AI drives. In P. Wang, B. Goertzel & S. Franklin (Eds.), *Proceedings of the first AGI conference, Volume 171, frontiers in artificial intelligence and applications*. IOS Press.

Penrose, R. (1989). *The Emperor's new mind: Concerning computers, minds and the laws of physics*. Oxford University Press.

Schmidhuber, J. (2009). Ultimate cognition à la Gödel. *Cognitive Computation,1*, 177–193.

Schwab, K. (2016). *The fourth industrial revolution*. World Economic Forum.

Silver, D., Hubert, T., Schrittwieser, J., et al. (2018). A general reinforcement learning algorithm that masters chess, shogi, and Go through self-play. *Science,362*, 1140–1144.

Tooby, J., & Cosmides, L. (1992). The psychological foundations of culture. In J. Barkow, L. Cosmides, & J. Tooby (Eds.), *The adapted mind: evolutionary psychology and the generation of culture*. Oxford University Press.

WID.world. (2018). *World inequality report 2018*. World Inequality Lab.

Yudkowsky, E. (2008). Artificial intelligence as a positive and negative factor in global risk. In N. Bostrom & M. M. Ćirković (Eds.), *Global catastrophic risks* (pp. 308–345). Oxford University Press.

Yudkowsky, E. (2013). *Intelligence explosion microeconomics* (Technical report 2013-1). Machine Intelligence Research Institute.

The manufacturer's authorised representative in the EU is Springer Nature Customer Service Centre GmbH, Europaplatz 3, 69115 Heidelberg, Germany. If you have any concerns regarding our products, please contact ProductSafety@springernature.com

Printed and bound by CPI Group (UK) Ltd, Croydon, CR0 4YY

30/09/2026

02225239-0002